Julius Krebs

The
VS Encyclopedia

Roland VS-2480 VS-2400 VS-2000

Translation by Beth Martin

PlancTon Verlag Berlin

All product names and trademarks mentioned in this book are subject to patent and legal goods protection. Despite highest care, mistakes cannot be excluded. For this reason, the publisher and author are not liable for incorrect instructions and their consequences, nor do they assume judicial responsibility.

Copyright © 2006 by PlancTon Verlag, Berlin
The author and publisher would like to thank Roland Japan, Roland Germany and Roland CE for their help and the pictures used in this document.
Cover design: PlancTon
ISBN 3-00-018103-2
2nd Edition

All rights reserved. No part of this book may be reproduced, processed, duplicated or distributed in any form without written permission of PlancTon Verlag.

Preface

Roland's workstations provide a host of recording, editing, mixing, and mastering functions, and even allow the burning of personal CDs. Understanding the complex processes, parameters, and operations involved, or indeed deciding when to use what, isn't always easy. Making the right decisions at the right time also requires some insight into the underlying technological concepts.

Hence this encyclopedia for the working musician and engineer, which covers all topics in alphabetical order. It is a hands-on guide, which provides countless solutions and tips that should allow even the most intricate issues to be solved in no time at all.

Whereas in an owner's manual, index entries usually refer to several chapters and/or paragraphs, in this encyclopedia, one entry covers all that is needed.

All VS functions described in this book are flagged with a "relevance gauge" that should allow the importance of the concept in question to be determined within the overall VS menu structure.
Be aware that, while most parts assume that the reader has a basic or fair knowledge of the underlying concepts, others are geared to experts.

A theme oriented "Guide" outlines the keywords of the most important
studio operations. These keywords are normally found on different pages due to the book being in alphabetical order. This means not only that choices can be preselected from the main themes before actually looking up a word, but also that structured learning is made easier.

See →Important parameters and operations (p. 248) for a general overview. Entries flagged with three bars refer to basic concepts all VS users need to be aware of.

In addition to step-by-step descriptions of important operations, the present publication contains countless hints that entry-level and experienced users alike will appreciate during their work with their Digital Studio Workstation.

Conventions

Entries:
All entries are listed in alphabetical order, which enables quick location of the information being searched for. This book focuses not only on operations, parameters, applications and the VS-series' controls, but also discusses fundamental audio and studio concepts.

Subsidiary entries often refer to their related main entries (Threshold →Compressor), while less frequently-used terms usually refer the reader to the more generic terms. Even for those less common terms, a brief explanation aids the decision of whether jumping to the "regular" term makes sense.

Entries that refer to complex subjects should be considered portals and road signs to help as a guide through all related concepts (example: Edit: →Region Edit (p. 188), →Phrase Edit (p. 150) and →Region Arrange (p. 184)).

Relevance Indications
Entries referring to VS terms are indicated with bars at the right. There are three relevance levels that should help to decide whether the information is really needed in a given situation:

Fundamental Concepts   ❚❚❚
Such parameters or functions are indispensable for working with the device.
Whilst easier and faster approaches may exist, operations flagged in this way allow the desired results to be quickly attained.

Advanced Use   ❚❚
These parameters or concepts provide a faster and easier way of getting things done during recording and mixing sessions. These two bars also indicate functions, which require more than basic studio knowledge for application.

Expert Use   ❚
Entries indicated in this way usually describe rarely-used operations, but may also refer to default settings. Usually, such concepts are only useful for professional users.

Abbreviations
1. The entries of the present book are usually abbreviated (only the first character is used). Example: Exciter: An *E.* generates…
2. The various VS models discussed in this book are abbreviated as follow:
   VS: Refers to all models (VS-2480, VS-2400, VS-2000)
   VS-24xx: Refers to the VS-2480 and VS-2400
   VS-2480:   Only refers to the VS-2480

Cross-References
Cross-references are marked by means of an arrow (e.g. →Record Monitor). They usually indicate where related subjects to be aware of, or additional useful information are to be found.

Capitalization
The names of buttons (keys) and knobs on the front panel are printed as they appear on the device. "EZ ROUTING" therefore refers to the VS's EZ ROUTING button.

Alphabetical Order
All entries are listed alphabetically according to the following system:

Spaces
Commas
Periods
Colons
Letters
Hyphens (where applicable)

About the author:

Julius Krebs, a graduate engineer, musician and composer, has been working on and for Roland pro audio products for many years. A hard disk recording specialist, he also teaches product seminars and conducts training sessions related to studio applications. In addition to his Roland assignments, he works as a free-lance producer, recording engineer and translator.

Guide

Recording (p. 176)

- →Project New (p. 168)
    - →Sample Rate (p. 199)
    - →Record Mode (p. 175)
- →Analog Inputs (p. 15)
    - →Phantom Power
    - →Line Level (p. 104)
    - →Guitar Hi-Z (p. 88)
- →Level Setting While Recording (p. 104)
    - →Level Meter (p. 103)
    - →Peak Indicator (p. 145)
    - →Input Clip (p. 96)
    - →INPUT Level Meter (p. 97)
    - →Clipping (p. 45)
    - →Recording of digital Signals (p. 181)
- →Patchbay (p. 144)
    - →EZ Routing (p. 74)
    - →Quick Routing (p. 171)
- →Mixer (p. 128)
    - →Input Mixer (p. 97)
    - →Track Mixer (p. 228)
    - →Fader Buttons (p. 77)
- →Headphone Mix (p. 92)
    - →PHONES (p. 146)
    - →Monitor Mix (p. 131)
    - →Record Monitor (p. 175)
- →Metronome (p. 120)
    - →Tempo Map (p. 223)
- →Tuner (p. 230)
- →Arming Tracks (p. 17)
    - →Record Standby (p. 176)
- →Punch In/Out (p. 170)
- →Recording Time (p. 183)
    - →Remaining Space (p. 192)
    - →Disk Memory Full! (p. 57)
    - →V-Track (p. 233)
- →Live Recording (p. 106)
- →Bouncing (p. 30)

Editing (p. 62)

- →Region Edit (p. 188)
    - →Phrase / Region Selection (p. 157)
    - →Region Copy (p. 186)
    - →Region Move (p. 191)
    - →Region Erase (p. 190)
    - →Region Cut (p. 188)
    - →Region Insert (p. 191)
    - →Region Compression/Expansion (p. 18.
    - →Track Import (p. 228)
    - →Track Export (p. 228)
- →Phrase Edit (p. 150)
    - →Phrase Copy (p. 147)
    - →Phrase Delete (p. 149)
    - →Phrase Move (p. 152)
    - →Phrase Split (p. 156)
    - →Phrase Trim (p. 157)
    - →Phrase Divide (p. 149)
    - →Phrase Export (p. 151)
    - →Phrase Level (p. 152)
    - →Phrase Name (p. 152)
    - →Phrase New (p. 152)
    - →Take (p. 221)
    - →Phrase Normalize (p. 153)
    - →Phrase Parameter (p. 154)
- →Region Arrange (p. 184)
    - →Marker (p. 109)
- →Phrase Sequence (p. 155)
    - →Phrase Pad (p. 153)
- Methods for →Editing (p. 62)
    - →Mouse Editing (p. 132)
    - →Numerical Editing (p. 136)
    - →Wheel/Button Editing (p. 243)
    - →Grid (p. 86)
    - →Scrub (p. 202)
    - →PREVIOUS/NEXT (p. 164)

Guide

Mixing (p. 125)
  →Channel (p. 43)
    →CH EDIT (p. 42)
    →Dynamics (p. 60)
    →Equalizer (p. 68)
    →AUX Send (p. 27)
    →Panorama (p. 141)
    →Channel Link (p. 43)
    →Group (p. 87)
    →Solo (p. 208)
    →Mute (p. 134)
    →Phase (p. 145)
    →V-Track (p. 233)
    →Link (p. 104)
    →MIX Switch (p. 128)
  →Effect (p. 62)
    →AUX Bus (p. 25)
    →Aux Master (p. 26)
    →FX Return Channel (p. 82)
    →External Effect Processors (p. 73)
    →Effect Economy (p. 65)
  Tools
    →Channel Strip (p. 44)
    →Analyzer (p. 16)
    →Knob/Fader Assign (p. 101)
    →Locator (p. 106)
    →Marker (p. 109)
    →Loop (p. 108)
    →Name Entry (p. 134)
    →Scene (p. 200)
    →Parameter Initialize (p. 142)
  →Automix (p. 19)
  View
    →CH EDIT (p. 42)
    →Multi Channel View (p. 130)
    →TR F/P (p. 226)
    →IN F/P (p. 96)
    →Channel View
    →Parameter View (p. 142)
    →WAVE DISP (p. 241)
  Mixdown
    →Stereo Mix (p. 211)
    →Surround (p. 213)
    →Live Mix (p. 105)

Effects (p. 62)
  →Send/Return Effect (p. 204)
    →Pre/Post (p. 162)
    →Aux Master (p. 26)
    →FX RTN (p. 84)
    →Loop Effect Assign (p. 108)
  →Insert (p. 98)
    →Bypass (p. 33)
  →Effect Algorithm View (p. 64)
    →Effect Patch (p. 65)
  Main Effects
    →Reverb (p. 193)
    →Delay (p. 53)
    →Chorus (p. 45)
    →Flanger (p. 79)
    →Phaser (p. 146)
    →Mastering Tool Kit (p. 117)
    →Speaker Modeling (p. 209)
    →MicModeling (p. 121)
    →Harmony (p. 89)
  →Plug-in (p. 159)
    →Auto Tune (p. 18)
    →Massenburg EQ (p. 110)
    →McDSP ChromeToneAmp
    →T-Racks Mastering (p. 220)
    →Soundblender (p. 209)
    →SRV Stereo Reverb (p. 210)
    →TCR 3000 Reverb (p. 222)
    →Urei 1176 Compressor (p. 231)
    →LA-2A (p. 102)
  →Effect Economy (p. 65)
    →Effects, recording (p. 67)
  →Effect Board (p. 64)
    →VS8F-2 (p. 241)
    →VS8F-3 (p. 241)

Synchronization (p. 217)
  →MTC (p. 133)
  →MIDI Clock (p. 123)
    →Sync Track (p. 215)
  →LANC (p. 102)
  →SMPTE (p. 207)

Guide

→Frame Rate (p. 80)
→Offset (p. 137)
→Current Time (p. 50)
→Tempo Map (p. 223)
→Time Display (p. 224)
→MIDI Sequencer (p. 125)
→Video Recorder Synchronization (p. 237)

Midi (p. 122)

→MIDI Connections (p. 123)
→MIDI Sequencer (p. 125)
→Control Changes (p. 49)
→Control Local Sw (p. 49)
→MIDI Clock (p. 123)
→MIDI OUT Sync Generation (p. 124)
→MTC (p. 133)
→MIDI Bulk Dump (p. 123)
→MIDI Effect Patch Select (p. 124)
→MIDI Mixer Settings (p. 124)
→MIDI Scene Change (p. 125)
→Mixer Control Type (p. 130)
→SMF (p. 207)
→System Exclusive Data (p. 218)

Mixer (p. 128)

→Input Mixer (p. 97)
→Track Mixer (p. 228)

Inputs
→Analog Inputs (p. 15)
→Digital Inputs and Outputs (p. 55))
→Patchbay (p. 144)

Outputs
→Analog Outputs (p. 15)
→Digital Inputs and Outputs (p. 55)
→Track Direct Out (p. 227)

Channels
→CH EDIT (p. 42)
→Fader (p. 77),
→Equalizer (p. 68)
→Dynamics (p. 60)
→AUX Bus (p. 25)
→Direct Path (p. 56)
→Insert (p. 98)

→Group (p. 87)
→FX Return Channel (p. 82)

Master Section (p. 112)
→Mix Bus (p. 128)
→Master Fader (p. 112)
→Output Assign (p. 139)
→Master Effect (p. 112)
→Master Insert with External Effect (p. 116)

Listening
→PHONES (p. 146)
→Headphone Mix (p. 92)
→Monitor Bus (p. 131)
→Mix Bus (p. 128)
→Record Bus (p. 174)
→Metronome (p. 120)

Metering

→Level Meter (p. 103)
→Peak Indicator (p. 145)
→Ext Level Meter (MB-24) (p. 72)

Routing

→EZ Routing (p. 74)
→Quick Routing Via Status/Select Buttons (p. 75)

Mastering (p. 113)

→Mastering Room (p. 115)
→Master Tracks (p. 118)
→Master Effect (p. 112)
→Mastering Tool Kit (p. 117)
→T-Racks Mastering (p. 220)
→Disc Image File (p. 57)

Tools
→Studio Monitors (p. 212)
→Speaker Modeling (p. 209)
→Analyzer (p. 16)
→Fundamentals (p. 82)

→Listening Level (p. 105)

→Loudness (p. 109)
→Mono Compatibility (p. 132)

Burning CD (p. 35)

→Audio CD (p. 17)
→CD Marker (p. 40)
→Disc at Once (p. 57)
→Track at Once (p. 226)
→CD Writer (p. 38)
→CDR (p. 41)
→Disc Image File (p. 57)
→Red Book Standard (p. 184)

Audio Export/Import

→Wave Import (p. 242)
→Track Export (p. 228)
→Phrase Export (p. 151)
→USB (p. 232)
→V-Fire (p. 233)
→Direct Outs (p. 56)
→Track Import (p. 228)

Harddisk

→Hard Disk (p. 89)
→Format (p. 80)
→Partition (p. 143)
→Fragmentation
→Project Optimize (p. 168)
→Safety in Operation (p. 199)
→Drive Check (p. 58)
→Switch off (p. 215)
→Shutdown (p. 206) pr

16 bit / 24 bit
The 16-bit →Word Length (resolution) with its 96dB dynamic range has long been the →Digital Audio Signal (→Audio CD) standard. Nowadays however, 24-bit systems with a dynamic range of 144dB are more common. The VS provides a →Record Mode parameter for selecting the new project's resolution (→Project New).

1-Bit Converter
These are also called →Delta Sigma Converters. They describe voltage differences between the current and previous samples.

3-Band Isolator
→Isolator, 3-Band

44.1 kHz
The →Sample Rate of 44.1kHz is the basis for creating an →Audio CD and can, excluding rare exceptions, be used for VS projects (→New Project). The VS-2000 only supports 44.1kHz.

96 kHz
The 96kHz →Sample Rate is only supported by the VS-24xx models. The VS-2000 only uses 44.1kHz to record audio data. Currently, 96kHz is only selected for →DVD-Audio and SACD (Super Audio CD). The advantage of 96kHz is not so much that it raises the audio data's frequency range to 48kHz, but rather that it enables work with the more efficient Anti-→Aliasing filters and →Dither algorithms for →A/D Conversion. Additionally better results can be attained while editing the audio data because of high-resolution equalizers* and effects. Attention should be paid to the following: More memory space is needed on the hard disk, the number of recording tracks is limited to 12 (→Recording) and some →Effect Patches are not available. Enhancement of the →Word Length, however, is of more importance to the signal quality.

> *The →Massenburg EQ uses 88~96kHz for its internal processing.*

+Insert
In the regular mode, the →Region Copy function normally overwrites the audio data at the →TO position. Choosing "+I", the audio data at the TO position is shifted right, towards the end of the Project, before new data is inserted. This automatic operation can also be performed separately with →Region Insert.

A

A.Gain
Auto Gain, →Compressor (p. 47)

A.PUNCH
With Auto Punch, recording processes which need to be repeated several times for musical reasons, can be automated. This function can also be used to avoid a complicated search for punch in/out points. The IN position determines the start point of the recording, while the OUT position determines the end point. These two positions can be set and edited in Stop and Play mode and can also be entered numerically.

VGA:	Right-click the →Measure Bar, set Auto Punch IN or OUT
VS:	Hold A.PUNCH and press IN or OUT

→Loop and Auto Punch (p. 108) can also be used advantageously together:

Activating Auto Punch

- While the project is stopped, press the A.PUNCH button or click the A.PUNCH symbol in the VGA display.
- Navigate just before the place where recording should start.
- Start the recording.
  When the IN position is reached, the VS automatically switches from playback to recording. When the OUT position is reached, the VS automatically switches back from recording to playback.

A/B Comparison

Editing The A.PUNCH Positions

1. By setting a new A.Punch point, the current point is overwritten. To delete an A.Punch point, hold CLEAR and click one of the Point icons on the VGA screen.
2. In the Utility menu's AUTO PUNCH/LOOP, the IN/OUT points can be entered numerically, changed or deleted. Here it is also possible to transfer the current time into the appropriate field and store it there:

VS:	SHIFT + A.PUNCH
VGA:	SHIFT + click [icon]
Utility:	Utility → APnch/Lop

Tip
If recording is not working, it is possible that Auto Punch is activated but that the current position lies outside the Auto Punch region.

A/B Comparison

By toggling between two source signals while keeping the same monitoring conditions, the sound quality, loudness, effect balance etc. of two mixes can be easily judged. This procedure is used both for →Mixing and →Mastering.

A/D Conversion

The analog/digital conversion translates analog signals into digital numerical values. The →Binary Numbers generated through this procedure are the basis for data storage and editing. A. analyzes the audio signals by taking samples. The number of samples taken is determined by the →Sample Rate. At 44.1kHz, there are 44.100 samples per second (see Oversampling). Each sample shows the applied voltage at its individual position, however, only finite numbers are used to show the voltage, depending on the →Word Length (bit rate). For example, 8 bits generate 256 steps, while 24 bits already provide more than 16 million steps. Binary codes can only show whole numbers. If a sample does not have an exact measurement value, it is assigned to the nearest stage. Rounding errors like this cause quantization noise, because values are permanently added to, or deducted from, the original voltage value. Due to this, digital signals never represent analog signals 100%. The transitions between individual amplitude values then are not continuous and the number of possible amplitude values is limited (see picture below):

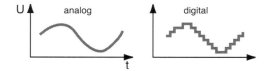

Fig.: Analog and digital signals

In the VS the →Word Length for the →New Project is defined by the →Record Mode. To achieve the highest dynamics possible, one of the two 24-bit formats should be selected ("MTP" or "M24"). For the →Sample Rate, select 44.1kHz to ensure compatibility with the Audio CD format. On the VS-2000 the Sample Rate is preset to 44.1kHz and cannot be changed.

A/D Converter

1. An A. converts analog signals to digital signals (→A/D Conversion). The audio quality is determined by the hardware used, the →Word Length, and the anti-aliasing filters. Depending on the number of bits being created, A.s are called "16-bit" or "24-bit converters". →Oversampling is usually used to suppress aliasing frequencies.

2. A.s are also available as external hardware units, mostly with a built-in pre amp or a full →Channel Strip. External high-quality A.s are often more expensive than the →DAW being used and can be utilized with the VS, especially if they are equipped with an analog →Compressor.

Tip
To expand the number of analog inputs, devices with a digital output such as DAT recorders or effect units can also be used as A/D converters (→Recording of digital Signals).

Absolute Time

During →Synchronization in slave mode, the incoming timecode can be displayed as an "absolute" value. To have an overview of special applications, it might be better to display the timecode with an offset (→Relative Time). The synchronization itself is not changed by this setting (VS-2000: Utility menu's "Display Parameter"):

Utility → Project → Time Display Format

Accessories

Depending on the desired practical application for the VS recorder, several options for accessories are available:

- →VGA Monitor (p. 236)
- →Keyboard PS/2 (p. 101)
- →Foot Switch (p. 79)
- Gig bag
- Mixer stand
- Pop Filter (suppress →Pop Noise (p. 162)
- →Meter Bridge (MB-24), p. 120 (not supported by the VS-2000)
- →DIF-AT24 (p. 54) (not on the VS-2000)
- →ADA-7000 (p. 13) (not on the VS-2000)
- →AE-7000 (p. 14) (not on the VS-2000)
- →VE-7000 (p. 236)
- →VS8F-3 (p. 241)
- →SI-80SP (p. 207)

Though the VS-2400 and VS-2000 could be installed in a 19" rack, the mounts required for this are not available.

ADA-7000

This 19" →R-Bus analog interface expands the number of analog inputs and outputs of a VS-24xx. Eight additional inputs with phantom power and eight additional outputs are made available. The A. also has →Word Clock IN/OUT connectors, which allow the use of several digitally connected units at once.

Fig.: Analog interface ADA-7000

VS-2400:
Only with the help of an A. are 16-track recordings with the VS-2400 made possible. The VS-2400 does not support →Dynamics and →Equalizer for Input channels 9~16.

VS-2480:
By using an A., 24 input channels* are supported. For this reason, the VS-2480 is perfect for →Live Mixes and →Live Recording (16 tracks).

* Inputs 9~16 are balanced phone connectors without phantom power.

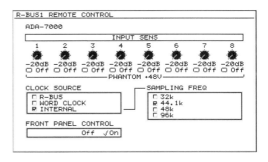

Fig.: R-Bus control page for the ADA-7000

ADAT

1. The ADAT recorder is a tape-based, digital 8-track recording system manufactured by the Alesis corporation. It uses VHS cassettes as recording media.

2. ADAT format: Developed by Alesis, an optical interface which uses Light Pipe to transmit audio data on eight channels simultaneously. To convert an ADAT signal to the →R-Bus format, the →DIF/AT or →DIF-AT24 interface box is required.

AE-7000

This digital interface A. broadens the number of digital connections of a VS-24xx. An additional eight inputs and eight outputs are available in the →AES/EBU format. Synchronization is controlled either by the incoming signal of input A, via R-Bus, or →Word Clock. The A. can be connected to the VS-2400 and VS-2480.

Fig.: Digital interface AE-7000

AES/EBU

A. is a digital 2-channel audio format with a maximum resolution of 24 bits. For data transmission a special 110Ω →Digital Cable (balanced) to be used with XLR connectors is recommended. The audio data format is the same as the →S/PDIF format. The difference is to be found in the user and control bits. The AES/EBU contains no copy protection (→SCMS). It is self clocking and requires no additional →Word Clock. With 5 volts, the signal amplitude is higher than S/PDIF signals and is therefore more stable. The VS-24xx models can send and receive AES/EBU signals with the help of the R-BUS interface →AE-7000.

Tip:
AES/EBU and S/P DIF are partly compatible with each other, which allows users to solder their own adapter cables. Usually the difference in voltage is no problem. It is nevertheless recommended to use a proper resistor when routing an AES/EBU signal to an S/P DIF connector.

AFL

"After-Fader Listen" is a special →Solo mode (Solo in place) which is also used in a VS. In this mode, the audio signal is outputted after the fader, equalizer and PAN knob. AFL is accomplished by muting all other channels.

VS-2480 and VS-2400:
While soloing mixer channels, the Effect Return channels are muted simultaneously. To listen to a Solo signal together with it's →Send/Return Effect, the FX-Return channel must be soloed as well.

VS-2000:
The PLAY/REC parameter →FX Rtn SOLO ENABLE allows the soloed channel to be heard together with its effect returns without further setting changes.

After Punch Out

A. determines the behavior of →Automix data after it has been corrected via punch in/out. With the setting "Return" for A. and a "Return Time" which is "> 0", the last edited values are faded to the next original event, in order to achieve a smooth transition. By choosing "Keep" instead of "Return", this function is deactivated and the result is a jump from newly recorded data values to existing data in the transition zone.

Utility →Automix Setup → "After Punch Out"

After Record

The A. parameter is used in the →Mastering Room to position the →Master Tracks that are generated. When creating several consecutive titles, the length of the blanks between songs can be automatically set. By selecting "Stay Here", the Master tracks and the Project tracks will be time aligned.

Algorithm

An →Effect's A. comprises the calculations to be performed for processing audio signals in a given way. This may be a basic algorithm (reverb, chorus, RSS, etc.), or a series of operations (i.e. processing blocks) of multi-effects (GuitarMulti, Vocal Multi, MasterToolKit, etc.). The VS currently provides 36 effect algorithms. Several variations (settings) of these are available as "Preset" and "User" →Effect Patches. The following preset, for instance, is

Analog Outputs

based on the "Microphone Modeling" (MM) algorithm: P124: [MM: 57 → 87]

Aliasing
This term refers to frequencies that are generated during an →A/D Conversion. Perceived as noise, and therefore undesirable, they are the result of unfiltered signal portions whose frequencies are higher than half the →Sample Rate.

All V-Tracks
Edit functions such as →Region Erase, →Region Insert and →Region Cut provide an "AllVTr" option that enables modifying all 16 →V-Tracks of the selected track in one go.

Application
The removal of a given passage from all 16 V-Tracks of one track simultaneously (or of several tracks). This allows the song structure to be changed not only for the currently selected Track but for all V-Tracks (if they are recorded). E.g. if several versions of a given part (i.e. lead vocals with different moods, phrasing, or notes) were recorded.

Analog Input Clip
→Peak Indicator (p. 145)

Analog Inputs
Analog Signals that are to be recorded (Mics, Synths etc.) are connected by a proper cable with the VS's analog input jacks. From here the signal is fed to a →Microphone Preamp. This preamp boosts the level to a value that allows the →A/D Converter to transform the analog signal into a digital one. The VS's analog input section features XLR and phone sockets. Both types are →Balanced. The XLR sockets are fitted with switchable →Phantom Power that is required for condenser microphones. The phantom power can be used independently for every XLR input (VS-2000: in pairs). The last input (IN 8 or IN 16) provides a socket for connecting high-impedance sources (electric guitars or bass guitars). This input needs to be activated using a dedicated switch (→Guitar Hi-Z), after which the regular input is bypassed.

	VS-2480	VS-2400	VS-2000
On Board	16	8	8
+ R-Bus	24	16	-

Tip:
As the VS models don't provide a switchable →Talkback input, a microphone can be connected to the last Input for talkback applications. Switching the talkback microphone on and off is then a matter of flipping the GUITAR switch.

Analog Outputs
The VS's analog outputs are used to transmit analog signals, particularly the stereo MASTER and MONITOR busses, or the headphone amps. The MONITOR OUTs are usually connected to the →Studio Monitors. The level of that bus can be set using the MONITOR knob. The MASTER OUTs, on the other hand, should be connected to an external recording device (DAT, tape deck, MiniDisk, etc.), if recording of the Mix doesn't take place in the VS. The →AUX OUT sockets can be used for separate headphone or monitoring mixes, for →External Effect Processors, or as →Direct Outs.

	VS-2480	VS-2400	VS-2000
Master	2	2	2
Monitor	2	2	2
AUX	4	4	2
Phones (L/R)	2	1*	1
R-Bus	16	8	–
Total**	24	16	6

* On the VS-2400, the PHONES outputs are hard-wired to the MONITOR bus and therefore can't be re-assigned.

** Excl. headphones

AnalogFlnger

With the exception of the VS-2000's AUX and MONITOR outputs, (RCA/phono, 0dBu) all analog output sockets of the VS are balanced and use the +4dBu output level.

Application:

- Connecting external (analog) recorders
- Monitor speakers
- External headphone amp
- →External Effect Processors
- Transmitting test signals
(not available on VS-2000)

AnalogFlnger

This →Effect simulates a vintage Roland SBF-325 →Flanger. It allows the creation of the typical moving and resonant sounds as well as the typical analog coloration produced by the original device.

Analog Flanger	
Effect patch:	P102
Connection:	Send/Return
Application:	Standard

Select "FL2" mode for →True Stereo operation. Use "FL3" for a more pronounced and intense effect.

AnalogPhaser

This →True Stereo effect recreates an analog-sounding 4- or 8-stage →Phaser (the number of stages is selectable). While it is often used for unobtrusive guitar chords or piano parts, it can also generate rather unique sounds with extreme cyclic level variations.

Analog Phaser	
Effect patch:	P099
Connection:	Send/Return
Application:	Standard

Analyzer

An A. (not available on the VS-2000) visualizes the frequency spectrum of a selected signal. The level read-out of the 31 →Third Octave Bands allows the visual examination of a sound and the simple comparison of signals or whole mixes.

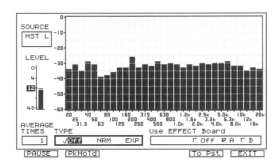

Fig.: Spectrum analyzer on the VS-24xx

Frequency bands are represented here by vertical bars, which show the current levels. Frequencies below 80Hz, which most monitor speakers cannot reproduce reliably, can often only be judged with the use of an A. The VS's A. requires the processing power of an entire →Effect Board (i.e. both processors of a VS8F-2), which needs to be assigned to the A. in the "USE Effect Board" field:

UTILITY → Osc/Analyz → F1 (Anlyzr)

The desired signal can be chosen with *SOURCE* (usually *MST L or R*)*. To achieve a smoother display, activate *AVERAGE* (for average analysis) and Peak Hold to maintain the highest level readouts. The display can be frozen using PAUSE.

Application:

- Optical checks of the frequency spectrum of individual signals or entire mixes (usually to see the sub-bass range).
- Measuring the frequency response of the listening room, etc.

* For comparisons, the →Generator/Oscillator can even be selected as the signal source.

Tip:

In VGA mode, the analyzer can be displayed independently on the VS's LCD screen irre-

spective of the currently selected VGA menu:
- Exit the analyzer using HOME.
- Select the F buttons of PAGE 3/3.
- Activate F4 (IDAnlr) and F6 (IDHold).

Arming Tracks
Arming a track determines the →Record Standby (p. 176) status.

Arrange
→Region Arrange can be used to change the order of song sections within a current Project. The Regions, defined by →Markers, affect all active tracks in the Playlist.

Assign
→Loop Effect Assign (p. 108)

Atapi
The CD drive built into the VS is called "Atapi" in the Project List. The "Advanced Technology Attachment Packet Interface" developed by the company Western Digital is a standard for controlling CD-ROM drives, CD writers and removable drives.

Attack
→Dynamics provide this function to delay the activation of a compressor/expander effect. This leaves the onset (attack portion) of the signals entirely unaffected.

Attenuator
Although an A. is normally used to cut signal's level, the A. in the VS also allows the boosting of a signal's level. This means that it can be used as an additional →Channel Strip tool for optimizing the signal level.

Setting range	[dB]
Attenuator	–42~+6
Channel fader	–100~+6

The attenuator is located in the signal path behind the →Dynamics processor and →Insert, but before the →Equalizer. This means it can't be used to increase the →Compressor's input level.

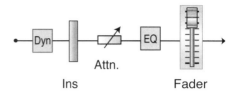

Fig.: Attenuator location in a channel's signal path

The EQ's attenuator shown in the LCD and the mixer channel's attenuator are one and the same.

CH Edit → Att

Application:
- Static level boost/reduction of certain channels for balancing reasons.
- Making up for marked level drops.
- Static correction ("offset") of already recorded Automix level values.

Tip:
The A. cannot be used to boost the level of the signal before it reaches the compressor. If low-level signals need processing with a compressor, consider increasing their levels using →Normalize.

Audio CD
The "Compact Disc Digital Audio" (CD-DA), a joint development of Philips and Sony, was introduced in 1982 as an optical carrier for storing digital stereo signals. The data is inserted into a track that spirals from the center (±20,000 rings totaling a length of roughly 6km), and which is covered by a reflecting aluminum layer. During playback, a laser lights the pits (holes) and lands (flat areas) inside the groove, which produces a diffusion or reflection of the laser beam. A photo sensor captures the number and pattern of reflections, which actually represent the digital code of the audio signals to be reproduced. During playback, the

Audio Data

revolution speed varies between 200 and 500RPM in order to maintain a constant reading speed. The "Red Book" defines the three sections each audio CD must contain: The "Lead-In" contains the table of contents (TOC) that lists the playing times of the tracks. The "Program Area" contains the audio data proper, while the "Lead-Out" indicates the end of the program area. The tracks are separated by "Track Pre-Gaps" that define the length of the silent portions (or "blanks"). Other constraints include the maximum number of tracks supported by a disc (99) and the minimum duration of all tracks (4 seconds). The maximum playing time is fixed at 80 min. The VS provides a →CD Burn function for the creation of CD-Rs that can be used as pre-masters for industrial CD duplication.

Audio Data

All A. handled and stored by the VS is called →Digital Audio Signals. It occupies the bulk of the device's storage capacity. The higher the →Word Length and →Sample Rate, the more space is required for the signal data. Other data types processed by the VS include →Automix events and →MIDI data as well as the →Events used for performing internal operations.

Auto Display

In the VS-2480, the multi-function PAN knobs (encoders) can be used to set parameter values. This function is called →Channel Strip. Activating the AUTO DISPLAY function automatically displays the →CH EDIT page associated with the parameter currently being edited.
The following settings are available:

→AUX Knob Auto Display (p. 26)
→Prm Knob Auto Display (p. 165)

Utility → Global Prm2 → XX Knob Auto Disp

Auto Gain

The →Compressor's parameter A. (Make Up) automatically boosts the level after gain reduction has occurred. By default, this parameter is set to ON, therefore the signal sounds louder when the compressor is activated.

Auto Locate

The VS-2480's →Phrase Sequence function automatically enables samples to be arranged in a rhythmic →Grid. This acts as a →Quantization for audio data.

Auto Punch

Aside from being able to manually activate →Punch In/Out for recording, this operation is automatically carried out by programming the IN and OUT time points via A. (→A.PUNCH (p. 11)).

Auto Tune

The A. →Plug-in from the company Antares corrects pitch deviations of monophonic signals in real-time based on the specified reference scale or MIDI entry. This plug-in requires the installation of a →VS8F-3 →Effect Board.

Auto Tune

Effect patch:	Plug-in
Connection:	Insert
Application:	Studio Standard

Fig.: Antares Auto-Tune VS (Plug-In)

The reference scale can be selected from among the common chromatic, major and minor scales, or from a series of historic and ethnic scales. Auto Tune can correct pitch deviations of up to one semi-tone in real time according to the closest "correct" note of the selected scale. For example: if the C major scale were selected, C# notes would be considered sharp (or flat) and therefore changed to either a C or a D (depending on which is closest to the original pitch). Though deviations in excess of a semi-tone are also corrected, such "corrections" usually lead to unexpected results.

After loading A. (→Plug-in) for an effects processor and assigning it to a channel as →Insert effect, press F4 (Edit) to call up the edit page.

1. Use "Input Type" to select the desired instrument or vocal range signal.
2. If the deviations don't exceed a semi-tone, set "Scale" to "Chromatic".
   Otherwise, specify the key and mode (major or minor).
3. Specify the song's "Key" (not necessary for "Chromatic").
4. "Retune" specifies how fast, in milliseconds, flat or sharp notes are corrected. Values between 10 and 50ms are usually perfect for vocals.
5. "Tracking" refers to the number of waveform cycles that are analyzed to determine the pitch of the incoming signal. Clean signals (without background noises) require fewer cycles, and thus a lower value. Unwanted results during pitch correction usually require an increase in "Tracking" value.

Display

The currently applied correction in cents* is shown in real-time – 1 semi-tone is 100 cents. Negative values refer to slightly sharp pitches with respect to the nearest scale note. Positive values indicate flat notes. "0" means that the pitch is perfect. Be aware that, depending on the "Retune" value, pitch corrections may be applied more slowly.

'Cher' Effect

Set Retune to"0" to apply instant pitch correction also eliminating natural vibrato. This creates the popular "Cher" effect.

Removing Or Forcing Notes

It is possible to not correct ("Bypass") or not output certain notes ("Remove"). To do so, click the appropriate column of notes that are to be excluded.

Remove:
"Remove" suspends the output of marked Notes. Any deviation in excess of a semi-tone from the intended pitch is then shifted to a more appropriate note. (Example: at one point in a C-major song, the singer overshoots and produces a pitch slightly above the "C#" (rather than a "C"). Normally, Auto Tune shifts that pitch to the nearest permissible note, which would be a "D" here. By setting "D" to "Remove", however, D is then excluded from being outputted. Consequently, Auto-Tune corrects to the nearest available note— the "C".

Bypass:
Activate "Bypass" for any note, which doesn't need correction. This is useful for parts that have good intonation and only one incorrect tone. To correct only that note, switch all other scale notes to "Bypass".

Specifying The Pitch Via MIDI

Forcing (corrected) notes from a scale can also be performed via MIDI. With "Target Notes via MIDI", the chromatic reference scale is selected. Firstly all notes are excluded by "Remove". Only notes played on a connected MIDI keyboard are outputted (and corrected for their pitch). Selecting another scale clears all selections in the "Remove" column.

Automix

1. Description

With A., the settings of various mix parameters can be stored and recalled. This will typically apply to the channel faders, but also pan, EQ, AUX sends, effect settings, surround motion (except on the VS-2000), parameters for →V-Link, etc. are easily automated. The VS distinguishes between the continuously varying settings of mix parameters (Dynamic Automix)

Automix

and the snapshots (→Total Recall), which represent a complete static setup. A. data is stored via →Events on separate tracks, independent of all audio data, and allows dedicated access for editing. For correction, the existing data can either be overwritten, modified graphically or with Micro Edit. Additionally, up to nine different mix versions can be stored per Project.

Note:
Audio and Automix data are recorded independently and therefore need to be edited separately in order to maintain their alignment in time. When editing audio passages using →Region Cut or →Region Insert etc., the same passages of Automix data must also be edited.

2. Dynamic Automix

Dynamic Automix allows the storing of continuous parameter changes for reproduction at any time. Usually, this would apply mainly to the fader movements.

2.1. Enabling Automix

The first step for storing or recalling Automix data, whether for dynamic data or for snapshots, is to enable the Automix mode. This can be done in one of the following ways:

2.1.1. VS Control Panel

Press AUTOMIX and the CH EDIT buttons either flash or light, depending on their current status (except on the VS-2000).

2.1.2. VGA display

Click the A button and the Playlist will show the current status of each channel in color.

2.2. Selecting Channels for Automix

Generally only the desired mixer channels are to be armed for Automix. In doing so, Faders and Pans can be automated without further settings. Channels to be automated can be selected as follows:

2.2.1. VS Control Panel

Press and hold AUTOMIX whilst pressing the required CH EDIT button several times (except VS-2000) to select the required status:

Status	VS-2400	VS-2480
Off	Yellow	Off
Record	Red	Flashes
Play	Green	Lights

2.2.2. VGA Display

In the Playlist, click the A button of the required channel. Clicking the ▲ button in the bottom row below the status indicators changes the status of all channels simultaneously.

2.2.3. Automix Setup Menu

SHIFT + AUTOMIX calls up the setup menu for both the VGA and LCD. The menu can also be activated via the Utility menu. This is described in the following section.

2.3. Selecting Mixer Parameters for Automix

The "Automix Setup" menu (see figure) allows the selection of the "Writing Parameters". Level and Pan are selected by default and therefore don't need to be activated. Use the dial or click the check box to select the desired parameter. Press YES to activate or deactivate the selected parameter.

2.4. Record Standby for Automix

In the same way as for audio recordings, the "Automix" function firstly needs to first be set to →Record Standby. Hold down the VS's AUTOMIX button while pressing REC. Alternatively, click the REC ● button on the VGA screen. The AUTOMIX button flashes red, while an "AUTOMIX REC" message flashes in the display.

2.5. Recording Automix

- Jump to the beginning of the song or the desired starting position and press PLAY.
- Any changes made to the selected parameter settings are recorded. Use the dedicated controls (faders, knobs, buttons, dial) or the mouse to perform the desired changes. (Example: to change a channel's level, move its fader, or move the cursor to FADER in →CH EDIT (or →TR F/P) and change the level using the mouse or dial.)

Automix

- Press STOP to stop recording (the indicator goes dark).

3. AUTOMIX SETUP Menu

Press SHIFT + AUTOMIX or use the Utility menu (LCD or VGA) to call up the "Automix Setup" page where the mix parameters and channels to be automated can be selected. This page also provides a host of parameter settings. Due to the default settings, only the desired channels must be armed for a standard mixer automation (faders and pans). (AUTOMIX STATUS). Channels selected for recording (WRITE) are indicated by means of black squares. Channels whose automation data is played back (READ) are displayed with a black frame (→Writing Parameter (p. 244)).

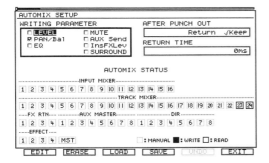

Fig.: The VS-2400's AUTOMIX SETUP menu

Note

The "Track Mixer", the section usually used for Automix applications, is assigned to the second row.

4. AUTOMIX PATTERN

Different automix variations can be saved under AUTOMIX PATTERN, of which nine different patterns can be stored per Project. Patterns can be generated by pressing F4 (SAVE) on the "AUTOMIX SETUP" page. They can also be named. Press F5 (LOAD) to recall a previously stored Automix pattern. Doing so overwrites the current Automix data.

Tip

VS parameter changes can also be recorded and recalled with a synchronized (→Synchronization) →MIDI sequencer.

5. Making Corrections

Recorded Automix data can be corrected by replacing old data, by overwriting, or by editing existing values. Use the special Automix UNDO function (see below) to discard the last changes, or ERASE (F2) to erase everything.

5.1. Overwriting

To overwrite an Automix passage, the Automix settings of the mixer channels and all parameters need to be set in exactly the same way as during the initial recording. Then start AUTOMIX REC. When the passage needing to be redone is reached, only the *first movement of the Fader (example)* rerecords the Automix data of this mixer channel. As long as no changes are performed, the original settings are preserved. Once recording is underway, additional changes can be performed with new fader movements. The last movement's value overwrites existing data until auto mix recording is deactivated. To stop overwriting press STOP, deactivate the Automix function (AUTOMIX + REC on the VS), or click the REC icon on the VGA screen. During overwriting, automix isn't active for the respective parameters and channels. It is also possible to use a manual →Punch In/Out for the mixer automation by activating/deactivating Automix.

The AUTOMIX SET UP's parameter "→After Punch Out" specifies the behavior of generated Automix data after a correction. Choosing "Return" causes the parameter to return to the previously recorded value. "Return Time" >0 sets the speed at which the original value is restored. "Keep" instead of "Return" generates an abrupt transition.

5.2. Editing Automix Data

Automix data can be edited both graphically and numerically. Editing existing values is often faster than overwriting them.

5.2.1. Graphic Editing

Automix data can be edited both in the LCD and on a VGA, which is made easy through the

Automix

graphic display. All Automix events, separated specially according to Parameters and Mixer channels, are recorded to dedicated "Event Tracks". On a VGA screen, such event tracks are superimposed on the →Waveform of the associated audio track, which helps to differentiate between the various tracks.

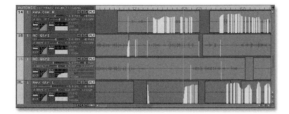

Fig.: Automix events in the VGA

Automix events are represented by vertical bars whose relative heights represent the parameter values.

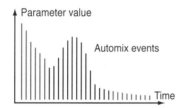

Fig.: Automix data

Graphic editing can be performed in the "AUTOMIX EDIT" menu.

LCD: SHIFT + AUTOMIX → F1 (EDIT)
VGA: Utility → Automix Edit

Start by selecting the desired parameter for viewing and editing.

Selecting an Automix Parameter

If only the level values need to be edited, there is no need to select this parameter, because it is selected by default

LCD: AUTOMIX SETUP →
F1 (Edit) → F1 (Target)

Fig.: Selecting the section and parameter (LCD)

VGA: AUTOMIX EDIT →
SECT: Section; PRM: parameter

Fig.: VGA's Section and Parameter Selection

Firstly specify the section containing the parameter needing to be automated. In most instances, this will be the Track Mixer, however, sections such as Input Mixer, FX Returns, Effects, MASTER, AUX busses and Direct Outs, can also be selected. The desired parameter can then be selected in the PRM field. The Automix edit functions operate similarly to the →Region Edit functions (Copy, Move, Erase and Insert). As the operational steps are the same, only how to copy data will be focused on here. This operation allows the duplication of desired dynamic changes so as to re-use them elsewhere (example: copying the automix passages of chorus 1 to other chorus sections). It is to be considered that audio and Automix data need to be copied separately. The procedure described under →Region Copy can also be applied to the Automix Copy function.

Stereo Tracks:

Level changes for →Linked mixer channels are recorded as OFFSET LEVEL events. Select this parameter for editing automix data, as well as for graphic and Micro editing (see below). "Level" and "Offset Level" events cannot be displayed simultaneously. The parameter OFFSET PAN is to be used in the same manner.

Automix

Automix Copy Quick Guide:
Select a parameter of the appropriate section (usually "Level") and specify the region to be copied. This region will be displayed inversely.

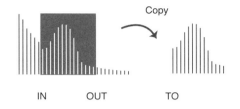

Fig.: Copying Automix data

Copying Using The Mouse:
- Draw a frame and release the mouse button.
- Click the region and keep holding the mouse button.
- Press and hold the SHIFT button.
- Drag the region to the desired destination.
- First release the mouse button, then SHIFT.

Numerically:
- Set the IN and OUT positions.
- Move the cursor to the source track and press YES.
- Set the TO position.
- Move the cursor to the destination track
- First press PAGE, then F1 (COPY).
- Press YES to copy the region*.

* The VS-2480 provides dedicated buttons for copy, move, insert, cut, compression/expansion and gradation.

Automatic Data Changes:
Data contained in the selected Automix region can be changed with the provided algorithms. These allow the "Compression" and "Expansion" of values, or to generate smooth transitions ("Gradation").

Comp/Exp:
The COMP./EXP. algorithm in/decreases the parameter values within the selected region. Changes via a parameter offset are ideal for fades whose last values turned out to be too low/high. For this, the "Shift" parameter allows the addition/subtraction of a set value to/from the data (offset). This preserves the original movement of the curve. If, however, the shape of the curve does need to be changed because the movement turns out to be too expressive or bland, modifications can be made with Parameter Threshold and Expand (>1 for compression or <1 for expansion.)

Gradation
G. automatically generates additional events between two existing Automix values. This can be used to smooth transitions, create fade ins/outs, etc.

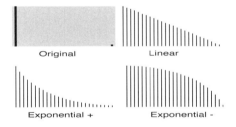

Fig.: Grading Automix data

These "interpolations" are performed according to one of three available curves: Linear, positive exponential, or negative exponential. "Exp+" is usually perceived by the human ear as the most natural movement.

5.2.2. Micro Edit
With M., Automix events can be controlled and changed in a table. This is especially useful for adding, (re)moving or changing individual events. This ensures that on/off parameters such as mutes, AUX Send switches, EQ switches or Pan modes will be produced at an exact time position.

Automix

LCD: Shift+Automix → F1 (Edit) → F4 (Micro)
VGA: Utility → Automix Edit → F4 (Micro)

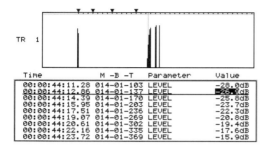

Fig.: Micro Edit for Automix data

Existing data is displayed both graphically and in a table. The Automix events of the selected parameter are listed in chronological order. To edit an event, it must first be selected. Within one line it is then possible to edit the value and the →Time Position of an event. As for graphic editing, regions can be defined and edited using the available functions (like →Region Edit, use PAGE for selecting). CREATE allows the addition and positioning of new events of the selected parameter. Without having to leave the page, CH INC and CH DEC can be used to jump to the same parameter of the next or previous channel.

Tip:

- For level corrections of whole tracks, simply the mixer channel's →Attenuator can be used.
- The level of an entire phrase can also be shifted using "Phrase Level" (→Phrase Parameter (p. 154)).

6. Automix Snapshot

"Snapshots" contain settings of selected parameter settings at the current time. During playback, the VS recalls the settings contained in the snapshot when it reaches that position. In addition to the Automix parameters that can be recorded individually, snapshots also memorize effect patch numbers. In a way, snapshots are similar to →Scenes, with the added convenience that snapshots can be recalled during playback. Scenes, on the other hand, contain Many more parameters than can be recorded by the Automix function.

6.1. Operation

Channel and parameter selections are performed in the same way as recording dynamic Automix changes (see above). The position where the snapshot should be recalled needs to be located (thus restoring the setting in effect at that precise position).
If necessary, change the settings, then proceed as follows:

VS: AUTOMIX + TAP

VGA: Click the [icon] icon

Application:

- Changing the effect patch number;
- "Static" automation instances that don't require transitions

Tip

If a Project contains several consecutive songs, consider inserting snapshots at the beginning of each title so as to save the individual, respective initial settings. This ensures the last settings of the previous song having no effect on the next one. (Otherwise, if the previous song ended with a fade out, the next song would start with a level of $-\infty$.)

6.2. Scene To Snapshot

Even though →Scenes are, in fact, snapshots of the current mixer settings, they can only be recalled while playback is stopped. This explains why Automix provides a Snapshot function. It is possible, however, to turn the settings of a Scene into a Snapshot by locating the desired position, stopping the VS and recalling the Scene in question. If a Snapshot is taken, it will be memorized with the Scene settings just loaded. This "Scene to Snapshot" trick also allows the use of identical mixer settings at various locations in the song (verses or choruses).

AUX Bus

6.3. Automating Effects

When effect processing power is limited (→Effect Economy), recalling various effect patches as the song progresses (again using Snapshots) dramatically expands the sonic possibilities. After enabling Automix (see above), in the "AUTOMIX SETUP" menu, select REC status for the corresponding effect (WRITE, black square). At the desired time, generate a Snapshot. In the VGA display, Utility → Automix Edit → Sect → Effect or F1 (TARGET) allows the selected effect page to be called up. To do this, click on the grey number field.

7. Automix Undo

The dynamic Automix provides a special Undo function that allows the condition prior to the last changes performed to be recalled. There is only one Undo level here (as opposed to the 999 available for the "regular" →Undo function). If the undone operation turns out to be better than expected, use REDO to restore it. It is possible to toggle between these two states until a new Automix operation is performed.

VGA Screen

Click the field (below the date) to activate UNDO. Click it again (→Redo) to restore the latest version.

'AUTOMIX SETUP' Menu

Press SHIFT + AUTOMIX on the VS or Utility → Automix Setup to call up the AUTOMIX SETUP menu*. The F5 button is assigned to the UNDO (or REDO) function.

  * →Fig.: The VS-2400's AUTOMIX SETUP menu (p. 21)

AUX Bus

A.s are used to bundle several signal sources and to transmit them to a common destination (→Bus). The VS's "Auxiliary Busses" are mainly used to send signals to internal* or external effects. To the user, they are only "visible" in their outputs being freely chosen (→Output Assign). Apart from a master bus fader, no further controls are provided for these busses.

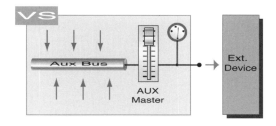

Fig.: Using an AUX bus to transmit signals to an external device

Application:

- Sending signals to the internal effects*
- Working with external effects
- Independent headphone mixes
- Sending separate mixes to external devices

  * VS-2000: →FX Bus
  Busses can be freely routed to any effects processor.

The arrows in the illustration above represent the channel signals transmitted to an AUX bus. The connection is established with the respective →AUX Sends.

Tip:

Individual signals can also be transmitted to internal and external devices using the →Direct Path function. This allows the AUX busses to be reserved for more universal applications.

AUX Master

Before the signals reach their destination they pass through an →Aux Master fader. This fader levels the output of the bus signal.

VS-2480:   IN 17~24 buttons (AUX1~8)
VS-2400:   AUX1~8 buttons (FX1~4)
VS-2000:   SHIFT + CH EDIT (MASTER)

AUX Level Meter

AUX level meters provide visual feedback of the corresponding bus levels. They are connected to the AUX Master faders (→Pre/Post) and

AUX Knob Auto Display

show the level of the signals being transmitted to an internal effect or an external device.

'Link' and 'Pre/Post' Routing

AUX busses can be →Linked (→Headphone Mix, etc.) to accommodate stereo signals. In addition, it is possible to specify for each mixer channel where the Send signal should be sourced:

Post Fader:
For "regular" applications (reverb, chorus, etc.).

Pre Fader:
For separate headphone mixes, etc.
These settings can be performed on the VGA screen (→Master Section) or in the LCD (→CH EDIT menus):

CH Edit (LCD):
Each mixer channel allows comprehensive "Link" settings to be established and to specify whether a signal should be sourced "Pre" or "Post". For pre/post move the cursor to "Pst" or "Pre" in the desired AUX field and press YES:

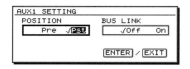

Fig.: AUX Bus Settings Using CH EDIT

Master Edit (VGA):
On the VGA monitor, the pre/post settings can be changed within each channel strip (→CH EDIT). The MASTER section is only to be called up for linking:

Mixer → Master Edit → F1 View

Change the "Off" setting of the desired AUX bus to "On".

Recording AUX Master signals
The bus signals transmitted to the selected destination can also be recorded (→Bouncing). To do so, use the EZ Routing environment to assign the AUX Master bus(ses) to the desired recording track.

Application:
- Recording the various channel send signal portions for subsequent effects processing.
- Recording →Surround signals.

* AUX and DIRECT Masters are located to the right of the Input Mixer on the "EZ Routing" page.

AUX Knob Auto Display

The VS-2480 allows the →AUX Send levels to be set using the multi-function "Pan" knobs (encoders). For a better overview, the corresponding →CH EDIT page can appear whenever activating an A.

Utility → Global Prm2 → Aux Knob Auto Disp

Aux Master

The AUX Master faders serve as a master fader for the →AUX Busses. This fader determines the volume of the bus signal, which is transmitted to an internal effect or an external device (→External Effect Processors, external headphone amp, →Surround monitor speakers, etc.

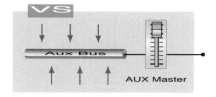

Fig.: Aux Master

Except for the VS-2000, all studio workstations provide faders for setting the AUX Master levels.

VS-2480:   Fader button IN 17~24 (AUX1~8)
VS-2400:   Button AUX1~8 (FX1~4)

The following methods are provided to adjust the Bus master levels in the Master menu:

VGA:	Mixer → Master Edit
VS-2400:	MASTER EDIT
VS-2480/2000:	SHIFT + MASTER

26

AUX Send

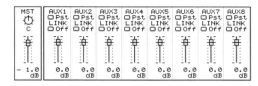

Fig.: AUX Master controls

On this page the outputting of signals of the mixer channels can be leveled*, set as →Pre/Post and →Linked.

> *In the →Surround environment, the AUX Master controls are used to set the levels of the Surround channels.

AUX OUT

These outputs can be used to transmit analog →AUX Bus and →Direct Path signals to →External Effect Processors. They can also be used for alternative →Monitor Mixes, as →Direct Outss, etc. (→Output Assign). They are configured as L/R pairs and need to be routed as such. The VS-24xx provides AUX A and B.

AUX Send

An Aux Send knob determines the level for the amount of the channel's signal, which is usually sent to an internal →Effect processor (→Send/Return Effect), to external destinations (Monitor mix), to →External Effect Processors etc. The assigned →AUX Bus is used as a medium for transmission.

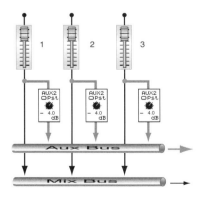

Fig.: AUX Sends (block diagram)

For a better overview, the illustration only shows three channels connected to one AUX bus. The VS provides eight buses that can be sourced either before (Pre) or after (Post) the fader (→Pre/Post), and also →Linked if necessary. In the illustration, the AUX Send signals are sourced Post fader. →CH EDIT settings still affect the signal transmitted to the AUX bus:

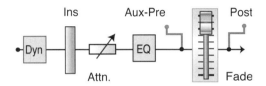

Fig.: Pre and Post Sourcing

1. LCD

Select the →CH EDIT page of the desired channel and move the cursor to the "Send Level" field. The default setting being "–∞", the desired value is selected using the dial or mouse. "*Off*" rather than "*Post*" means that the signal is not sent in the bus. This can be used for an →A/B Comparison of the processed/unprocessed signals. For an overview of the Send levels to *one* given AUX bus, there is a →Parameter View page that displays eight channels at a time.

CH EDIT → AUX Send field → F6 (PRM.V)

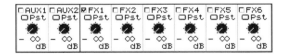

Fig.: AUX and FX Send in the LCD (VS-2000)

2. VGA screen

After selecting the →CH EDIT menu of the desired channel on the VGA, the Send levels can be set here in order to feed internal effects processors or external destinations. To do so, click the corresponding knob and move the mouse vertically, or use the cursor buttons and dial to set the desired value.

AUX SEND•PRM EDIT

Fig.: AUX Send on the VGA screen (CH EDIT)

Special VGA pages:
If necessary, all channel strips can be displayed on the VGA screen. Use the MIXER menu or PAGE 2/3 to select the →IN F/P (F2), →TR F/P (F3) and →MltChV (F4 not available on the VS-2000) pages. When clicking and vertically dragging the mouse in an AUX field, a blue horizontal bar appears that can be used to set the corresponding level.

3. Fader Controlled AUX Send Levels
The VS-24xx also allows the multi-function →Faders to be used to set AUX Send levels. The positions of the physical faders are useful for visual feedback of the Send levels specified for the various channels. To activate this "fader edit" function, press the button mentioned below (on the VS's operation panel) so that it flashes in red:

VS-2480:   KNOB/FDR ASSIGN (AUX SEND 1~8)
VS-2400:   Fader ASSIGN

To select an AUX bus, switch on the "Locator" button of the same number (Loc 1: Fader → Aux Send 1, Loc 2: Fader → Aux Send 2, etc.).

Example:
Select "Loc 1" to set the Fader controlled Send level of AUX 1 for each channel.

Fader 1	Fader 2	Fader 3	Fader 4	Fader xx
AUX 1 of CH 1	AUX 1 of CH 2	AUX 1 of CH 3	AUX 1 of CH 3	AUX 1 of CH xx

In this mode, the faders act as AUX 1 Send controls of the various mixer channels. By default, AUX 1 is routed to effects processor 1 (reverb). The fader, therefore, sets the reverb send level of the various channels without further adjustments.

Tip:
It is recommend to leave the "Fader-Edit" mode as soon as the required AUX Send levels have been adjusted. This avoids erratic settings due to changed fader assignments.

VS-2480:
In addition, the VS-2480 allows AUX Send levels to be set using the →PAN/AUX SEND Knobs (→Channel Strip). This may be simpler, because it eliminates confusion with the channel level function.

Utility → Global1 → Knob/Fdr Assign Sw

4. Automating AUX Sends
The levels of the various AUX Sends can be fully automated (→Automix). Use the "WRITING PARAMETER" field on the "SETUP" menu to select the "AUX SEND" entry →3. AUTOMIX SETUP Menu (p. 21).

AUX SEND•PRM EDIT
This button on the VS-2480 switches the →Channel Strip on/off. The color indicates the current status of the →PAN/AUX SEND Knobs. See the table below:

Function	Indicator	see
Panorama	Off	p. 141
AUX Send	Green	p. 27
Compressor/EQ	Red	p. 47/p. 68

AUX/DIR
The A. →Level Meters allow the levels of the →AUX Busses and →Direct Paths to be controlled both in the VGA and LCD screens. To select this, use the →Function Buttons or the corresponding fields on the VGA screen.

AUXF/P
Due to the limitations of the LCD display there is an additional page that displays the fader/pan settings of the →Aux Masters, →Direct Path Masters and the →FX Return Channels. This appears only when the LCD is used for operations (→Operation Display):

PAGE → F4 (AUXF/P)

To switch back to the →Level Meter screen, use PAGE in order to display the INPUT • IN Mix • TR Mix, etc., functions.
The VGA screen provides these parameters on the "→IN F/P" and "→TR F/P" pages.

Average Time
For the →Analyzer (not for the VS-2000), selecting a higher A. value makes the read-out smoother because the Analyzer uses average values for the various frequency bands.

B

Backup
This term refers to saving data to an external medium. VS Projects provide the following Backup options:
1. →CD Backup (p. 34)
2. →DVD Backup (p. 60) for VS-2480DVD
3. →USB (p. 232) for VS-2000: data transfer to a PC/Mac.
4. →Project Copy (p. 166) to a different →Partition of the internal hard disk. Although this is not an external media, it provides a certain amount of security.

Balance
The →Effect Return channels provide a BALANCE parameter that enables the effect's stereo placement within the bus mix to be specified. This parameter corresponds to the →Panorama function of regular mixer channels.

Balanced
A "balanced" connection provides two signal wires inside a common ground shield. The signal wires transmit the same signal with a 180° phase shift. The signal transmission is carried out only via the two wires while the shield is a HF protector. Given that the input section of the receiving device only monitors changes in voltage, any interference included in the signal is cancelled out. Balanced cables are used for microphones (→Microphone Level), professional line-level connections (+6dBu), and for the transmission of digital signals in the →AES/EBU format (→AE-7000). Unlike →Unbalanced cables, balanced ones (→Cable) may be hundreds of meters in length.

Band Eliminate Filter
Only the VS-2480 provides a BEF in addition to the normal →Equalizer filters. The BEF works in the same way as a peaking filter (cut only) with an extremely narrow frequency band. Its use as BEF (or "band reject filter") may come in handy to eliminate ground loops, other noises within a given frequency band, or also for suppressing certain tones (→Fundamentals). The frequency and bandwidth ("→Q", up to 16) can be set.

Bandpass
Only the VS-2480 provides a band-pass characteristic in the additional filter (→Equalizer). A combination of high-pass and low-pass filters, it cuts out the frequencies above and below the selected center frequency. Its main use is to create "frequency windows" during a →Mix.

Tip
As a substitute, a band-pass filter with a limited slope can be constructed using the channel EQ's shelving filters. Set their Gain values to "–15dB". The center frequency is determined by the two cutoff settings. (Example: "Band-Pass" with a center frequency of 1kHz: Low= 100Hz, High= 10kHz).

Bandwidth

Bandwidth
The "bandwidth" refers to the width of the effected area of a peaking filter in a parametric →Equalizer. This is not the same as the VS's "→Q" parameters. Its slope is expressed in →Octaves. See the following table:

Q	0.4	0.67	1.41	2.87	4.8	7.21	14.4	16
BW	3	2	1	0.5	0.3	0.2	0.1	0.09

Bank Select
The Preset and User Patches of the VS's effects processors can also be selected by subsequently transmitting the 0 and 32 →MIDI Control Changes from an external device (→Effect).

Battery
When the message: " !!! Warning !!! Clock/Calendar Backup Battery low " appears, the internal battery for powering time keeping features and temporary parameter values must be replaced.

Beat
1. The "Beat" usually refers to the number of quarter notes (e.g. 4/4 or 3/4 bar) or eighth notes (e.g. 6/8) in a bar.
2. In the VS's →Tempo Map, the time signature is called "Beat".

BEF
→Band Eliminate Filter (p. 29)

Binary Number
B.s refer to a system that simply uses the digits "0" and "1" (to indicate the "Off" or "On" state, →Bit). Digital technology is based solely on binary numbers.

Bit
A "bit" ("binary digit") is the smallest information unit used by digital technology. It can only have the logical switching states "0" or "1", which correspond to predefined voltages (0= ±0V; 1= ±5V). Digital signals are encoded as a series of zeroes and ones, which has the advantage that their significant differences in voltage can even be transmitted down bad signal connections without any mistakes. The number of bits used within a system is referred to as the "→Bit Rate".

Bit Rate
The "bit rate" (word length, resolution, or →Record Mode on the VS) refers to the number of →Bits used to encode analog signals. It determines the number of volume levels, and thus the →Dynamic of a signal. Though the CD standard still relies on a 16-bit resolution, professional audio applications nowadays use 24 bits (→).

Boost
B. refers to increasing the level (→Gain) of an →Equalizer filter band. Decreasing the level is called "cut".

Boot
Each time the workstation is switched on, the operation system is loaded and a number of setting routines are performed. This is called "booting". The VS firstly checks the internal hard drive* and the CD burner ("Setup Drive" or "Setup IDE"). If a System CD was inserted, a →System Update can be performed. The second step is to scan for effect expansion boards ("Setup VS8F-2, VS8F-3"), while the next step restores the last used Project. When that is done, the VS is ready for operation.

VS-2480: also external SCSI drives

Bouncing
B. ("submix", "ping-pong") refers to the internal re-recording of already recorded tracks to another track. A Phrase generated in this way is a physical entity that uses an additional amount of disk space. The destination track can be specified as a mono or stereo track. The stereo track preserves the original L/R settings. Track signals, effect return signals of both internal and external effects processors, and also

Brickwall Filter

AUX Send and Direct Path signals can all be bounced. All Settings go into the destination track in the form of data.

Application:
- Whenever additional tracks are needed.
- For reasons of clarity (e.g. combining backing vocal, drums, etc.).
- To combine several takes into a single track (e.g. the beginning of a guitar solo on track 6 with the ending from track 7).
- To record →FX RTNsignals (printing effects (→Effect Economy (p. 65)).
- To create stereo mixes →Master Tracks (p. 118)
- Recording →Surround (p. 213) channels.
- Creating →Subgroup (p. 212)

Operation
In the following illustration, tracks 1, 16 and 24 of the Track Mixer as well as →FX Return Channel 4 are routed to tracks 9/10.
Position the cursor over the desired source channel in the Track Mixer row. Using the dial, a connection with the recording tracks located above (→Routing) is easily established. While bouncing to armed destination tracks, the VS disconnects all source tracks from the →Mix Bus. It is only possible to hear the source tracks, via the destination track channels (→Record Monitor).

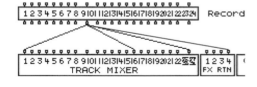

Fig.: Bouncing

- In the "→EZ Routing" page, routings are to be established from the Track Mixer (source tracks) to the desired RECORDING TRACKs (destination) via virtual patch cords.
- Simply record-enable the destination track(s) (its STATUS indicator flashes red) and start recording.
- Monitoring is established via the destination track's mixer channel.

Recording Effects
Whenever →Effect Economy is an issue, the →Effect Return signals can be recorded on an additional track (→Effects, recording).

EZ Routing Template:
The "P01: BOUNCING" template of the "EZ Routing" is provided for bouncing applications. It connects all playback track channels (except the last 2 tracks) and the effect returns to the last two tracks.

EZ-ROUTING → F5 Load → P01: Bouncing

AUX and DIR Masters: The "EZ Routing" environment even allows the recording of AUX Master signals (→AUX Bus) to the desired track(s). To do so, establish a connection between the AUX (FX) Master and the desired channel.

BPM
Refers to "Beats Per Minute" and hence to the song tempo. "♩ = 60" represents 60 quarter notes per minute, while "BPM= 120" refers to 120 quarter notes per minute. The VS allows the entering of tempo values in its →Tempo Map. For more flexible tempo changes, consider importing the tempo of a →MIDI Sequencer song into the →Sync Track.

Brickwall Filter
B. are steep rejection filters that suppress undesirable frequencies contained in an audio signal. The VS provides an →Isolator, 3-Band effect for such applications with fixed frequencies. The VS-2480 offers even more flexible options with a combination of EQ and filter.

VS-2480
Layering the filter and EQ bands on the VS-2480 enables extremely steep rejection filters.

Bulk Dump

The table below lists the settings to use for high- and low-cut filters.

Type	Filter	EQ (Q = 2)		
		Shelving	Low-Mid	Hi-Mid
Low cut at 100Hz	BEF 100Hz	Low: –15 dB / 20 Hz	–15 dB / 30 Hz	–15 dB / 40 Hz
High cut at 10kHz	BEF 10kHz	High: –15 dB / 20 KHz	–15 dB / 20 KHz	–15 dB / 20 KHz

Table: Configuring brick wall filters on the VS-2480

The following figure shows the response curve of a "Brickwall filter", created with the VS-2480's onboard filters:

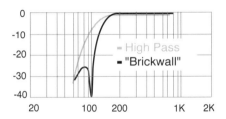

Fig.: "Brickwall" filter as 100Hz low-cut

Bulk Dump

→Scenes, →EZ Routing templates and User →Effect patches can be transmitted to an external sequencer (or another VS of the same series) via →MIDI. At the moment, this approach is the only way to use Scenes across Projects.

Application:

- External storage of data (Safety copies).
- Using the same mixer settings (Scenes) in various Projects.
- Converting data into →SMF and sending it to other VS users via e-mail.

Transmitting Settings To An Sequencer

Connect the VS's MIDI OUT socket to the →Sequencer's MIDI IN port. Next, jump to the "MIDI" page and select the "BULK DUMP" menu. Make the following settings:

UTILITY → MIDI Prm → SysEx. Tx Sw = On
→ F5 (BLK Dmp)

Page: 'MIDI BULK DUMP' (e.g. 'Scene')

1. Select the desired Scene number in the "BULK Tx TARGET" field (or specify "All" for all Scenes).
2. Click "On" in the "Scene" row of the "BULK Tx Sw" field.
3. Set the MIDI sequencer to record →System Exclusive Data and start recording.
4. Press F1 BULK Tx.

Receiving Data From A MIDI Sequencer

Connect the VS's MIDI IN port to the sequencer's MIDI OUT socket. Next, jump to the "MIDI" page and set the "SysEx Tx Sw" to "On".

Utility → Midi Prm → SysEx. Rx Sw = On
→ F5 (BLK Dmp)

1. Press F2 (BULK Rx) in the "BULK DUMP" menu. The VS now waits for incoming MIDI data.
2. Start playback of the external sequencer.
3. When all data has been received, the display asks whether it is OK to overwrite the original memory locations.

The Scene data is always stored to the memory locations that contained it when settings were dumped. Scene settings most likely to be used at a later stage should be copied to a different place beforehand.

User effect settings and routings can be archived in the same way.

Burning CDs

The VS's built-in CD writer enables the burning of CD-Rs. This function is available for
→CD Burn (p. 35)
→CD Backup (p. 34)
→Wave Export (p. 242)
→Wave Import (p. 242).

Bus

'Bus' is a typical mixer term. Busses bring feeds from the various channels together to be routed to a common destination. In this way the signals of all mixer channels, Effect Returns and subgroups are sent to the →Mix Bus, and from there to the MASTER section. By contrast, an →AUX Bus transmits a separate "mix" to internal or external destinations, e.g. the internal effects processors (on the VS-2000 called "→FX Buses"), or to a headphone amplifier. The →Monitor Bus is used for listening purposes and can be routed to other busses. During recording, it is possible to listen to the signals being recorded with the →Record Bus.

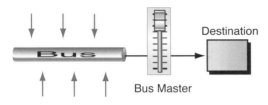

Fig.: Busses and Bus Masters

Unlike a bus, a →Direct Path transmits only one signal to the desired destination.

Bypass

The BYPASS switch temporarily deactivates the effect to which it is assigned. When "On", the signal runs over another path and no longer in the corresponding effects processor. The BYPASS function is useful for comparing an unprocessed ("dry") signal with an processed ("wet") version.

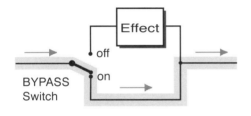

Fig.: Effect BYPASS switch

The BYPASS function of each effect can be activated/switched off both on the "EFFECT VIEW" and on the corresponding Edit pages.

Bypass ON = No effect

Note
Due to the latency induced by certain →Plug-ins, the BYPASS setting may produce a phasing effect in the →Send/Return Effect.

C

Cable

Most recording applications require the use of different cable types for the devices being used. Analog microphone and line-level signals are transmitted along balanced or unbalanced shielded cables. Speaker connections on the other hand, don't require shielded cables. Due to high signal voltages, electrical interference here usually doesn't affect the signal quality. Be sure, though, to use an appropriate wire width. Digital connections require special →Digital Cables capable of transmitting high-frequency (several MHz) analog signals with low characteristic wave impedance.

Unbalanced
This cable type contains one signal wire and a shield. The shield offers sufficient protection from low-level interference and constitutes the 0V potency. Unbalanced cables are used for high-level signals (→Line Level, e.g. CD players, synthesizers, samplers, etc.) with a maximum lengths of approximately 10m. The digital coaxial sockets (→S/PDIF) are also unbalanced.

Balanced
A "balanced" connection provides two signal wires inside a common ground shield. The signal wires transmit the same signal with a 180° phase shift. This means that both wires are required to transmit information, while the ground wire acts as shield. Given that the input

CC

section of the receiving device only monitors changes in voltage, any interference included in the signal is cancelled out. Balanced cables are used for microphones (→Microphone Level), professional line-level connections (+6dBu), and for the transmission of digital signals in the →AES/EBU format (→AE-7000). Balanced cables may be hundreds of meters long.

CC

Control Changes represent a special kind of MIDI data. The VS uses them to control external MIDI devices (→MIDI) or to transmit and receive mixer settings.

CD

The VS provides the following CD functions for the internal CD drive*:

1. Creating →Audio CDs that can be played back on regular CD players (→CD Burn (p. 35)).
2. Playing back commercial CDs and unfinalized CD-Rs.
   (→CD Player (p. 40)).
3. Ripping CD tracks to the VS's hard disk (→CD Capture (p. 39)).
4. Archiving Project data, In doing so, all V-Tracks and mixer settings can be stored and recalled (→CD Backup, →CD-RW).
5. Importing and exporting WAVE data to/from tracks and phrases. Commercial WAV data can then be added to recordings made on the VS. Conversely, VS tracks can be exported to other digital audio workstations (→Wave Import (p. 242), →Wave Export (p. 242)).

* *The first VS-2480 model has no internal CD drive. For this model, an SCSI drive needs to be connected.*

CD Audio

→*Audio CD (p. 17)*

CD Backup

The VS's →Projects can be saved to →CD-R or →CD-RW, as a backup (→Safety in Operation) and to use projects in other VS models (→Project Export). The archives created contain audio and recorder data (→Takes) as well as the respective mixer, MIDI, synchronization, Automix and Global settings. Depending on the data size, one CD-R can accommodate several Projects. In other instances, several CDs may be needed for archiving just a single Project. A Project's size is displayed next to its name in the Project List. This indication helps to calculate the number of CD-Rs required to archive the Project (most CD-Rs provide a storage capacity of 800MB). It is impossible to write to a CD-R/CD-RW that already contains data.

PROJECT → Mark → Page → Backup

Note
The Backup function is only available for Projects from the currently selected →Partition.

1. Selecting a Project
Calling up the Project List:

SHIFT + F1 or VGA menu bar → PROJECT

Select the Project(s) needing to be archived via wheel or clicking and mark it/them using MARK (F6) or YES. Selected Projects are indicated by a check mark.

2. 'BACKUP' Menu
Various settings can be found in the Project List's "Backup" menu.

PAGE (repeatedly) → F1 (BACKUP)

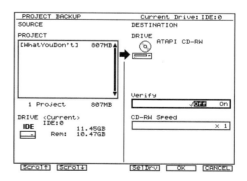

Fig.: 'BACKUP' Menu

Verify: After burning data, the VS compares the Project data with the data on the CD-R and displays any possible detected errors. To ensure the safety of the work, choose "ON".

CD-RW Speed: To start with, the speed setting of "MAX" should be chosen. Only when an error message is displayed should this setting be decreased.

Settings need to be confirmed with YES. The following "Save Current" should also be confirmed with YES if the Project contains changes not yet saved. *.

Several CD-Rs

For large Project files (>800MB), several blank CD-Rs will be necessary. During the Backup operation, the VS will stop whenever the current CD is full and display "Insert Disc x". It is recommended to label the CDs for subsequent Restore operations.

> * *To work with a →CD-RW that already contains data, it must first be erased.*
> *(using the dialog window that appears).*

The VS then starts the Backup routine and finally ejects the CD-R.

The operation →CD Recover is used to reload previously archived data back onto the VS.

CD Burn

To turn a CD-R into a regular →Audio CD, the Project needs to have a →Sample Rate of 44.1kHz. The →Record Mode, on the other hand, can be chosen freely (→Dither, →Project New). Two V-Tracks act as the source. Normally the Mastering Tracks, recorded in the →Mastering Room, will be used. (It is possible, however, to select any two tracks.) Only the audio data on the hard disk, without the mixer, effects and playback setting, is written to an audio CD. In the →Master Tracks, however, the audio data is already processed by the mixer and effect settings. The various tracks to be written to CD can either be written simultaneously (Disc At Once), which requires rather lengthy preparations, or simply one after another in separate sessions. Though this "Track At Once" (or TAO) approach doesn't completely conform to the →Red Book Standard, the duplication companies nevertheless accept such multi-session CD-Rs.

1. Preparing The Source Tracks

The source tracks usually require some form of editing (truncating/trimming, for instance) before burning them. If several tracks are to be burned at once, additional operations such as copy, position, sequence and add CD markers need to be carried out. Note that with the exception of the VS-2000 the Mastering tracks are edited in a different location:

1.1. Editing Location

1.1.1. Mastering Room

The →Master Tracks generated in the →Mastering Room can be edited right there, irrespective of whether they use the →CDR* or Project format. Only the VS-2000, however, allows for →Editing the Mastering tracks directly. The VS-24xx models require that the V-Tracks corresponding to the Mastering tracks are edited.

> * *Recognized by an "*" marked before the track number. Generated in the Mastering room with "CDR Rec Mode" set to "ON"*

Operation

- Activate the Mastering Room and select the "PLAY" status.
- Reveal the Mastering Tracks* by →Scrolling vertically (Playlist, below track 24).
- Link Track Mixer channels 23/24 (→Link).
- Select the →V-Tracks that contain the Mastering track data (default: 23/24–16). Their length and position are identical to those of the Mastering tracks.
- Edit the corresponding tracks (see below). The Mastering tracks change simultaneously.

VS-2000:

The Master Tracks are permanently shown below Tracks 17/18. In contrast to the VS-24xx models, they can be edited here directly.

CD Burn

1.1.2. Active Tracks

Mastering tracks in the regular Project format (mostly MTP, not CDR!) can be edited like any other track - there is no need to enter the Mastering Room. The V-Tracks in tracks 23/24, which were used as the Master Tracks, must be selected (in the VS-24xx usually Track 23/24-16).

1.2. Editing The Source Tracks

1.2.1. Clean Up

For a clean start of a Track, Pauses and noises etc. must be removed. In the same way, the end of the track must also be edited.

 The easiest way to do so is by using "→Trim In/Out" (→Phrase Mode: position the mouse on the edge of a Phrase (the bar) and click/drag the mouse). Or draw a frame on top of the area that is to be removed and delete that region (→Region Mode is selected by default).

1.2.2. Move To '00:00:00:00'

 Regular audio CDs require that the first track be located at the very beginning. Move this track to the beginning of the Project. Activate →Phrase Mode, click one of the two phrases and drag it to the left ("00:00:00:00"). Alternatively, and to move all subsequent tracks together, →Region Cut can be used to remove the area between "0:0:0:0" and the actual beginning of the track. Subsequent Phrases are moved by the same amount.

1.2.3. Using Several Tracks Within A Project

If not already done, the future CD tracks should be copied to the desired stereo track (→Region Copy). Position the CD tracks in the desired sequence and add the required blanks. Doing this eliminates yet another editing step.

- Trim each track to the desired length.
- →Phrase Mode allows titles* to be moved with the mouse in one step, which in turn enables the desired sequence and pauses to be set.

* *Tracks, which are combined from various phrases need to moved via →Region Copy.*

If necessary, use →Track Import to add the Mastering tracks of other Projects to the CD sequence.

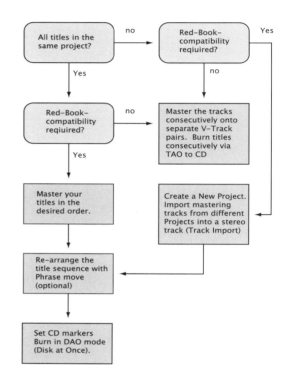

Fig.: Combining tracks for a CD

2. CD Markers

CD players allow following or previous tracks to be skipped or selected by specifying their numbers. With the VS this function becomes possible thanks to "CD markers". The VS generates the markers automatically for individual tracks (TAO), but also allows them to be set manually.

2.1. Multi-Session CDs

For CDs that only contain one track (or several tracks written in subsequent sessions), there is no need to set CD markers. The VS generates them automatically and inserts 2-second pauses between tracks.

CD Burn

Multi-session tracks

Write method:	Track at Once
CD Markers:	Automatic

Select →Track at Once as the WRITE METHOD for burning the CD.

2.2. CDs Containing Several Tracks

In the Master Tracks, a CD marker must be inserted at the beginning of each song in order to choose this track on any CD player (see below).

2.3. Generating CD Markers

- All operations concerning CD →Markers can be performed in the Mastering Room.
- Position the track cursor on one of the source tracks in the →Playlist.
- Press ZERO to move to 00:00:00:00.
- Press the →NEXT* button twice to locate to the beginning of the *second* CD track. Usually the title comprises of one Phrase, otherwise press the button several times. (The index marker of the first track is inserted automatically.)
- Hold down the CD-RW/MASTERING button while pressing TAP to insert a CD track marker at the →Position Line.
- Check the settings using →PREVIOUS/NEXT.

The NEXT/PREVIOUS buttons jump to the nearest markers or Phrase ends. This depends on the →Previous/Next Switch setting.

Alternatively, CD track markers may be generated in the Mastering Room by right-clicking in the measure bar (just above the Playlist).

VGA:	Right-click the Measure Bar → CD Marker
VS:	CD/RW + TAP
VGA+VS:	CD/RW + click TAP

Tip:
After creating CD markers in the VS, their numbers do not yet coincide with the corresponding track numbers on the CD. To work around this, create two normal markers in the first title. This operation has no effect on the content of the CD track itself.

2.4. Editing CD Markers

CD markers can easily be moved using the mouse or deleted by holding CLEAR and clicking on a marker. All →Markers can be deleted via SHIFT + CLEAR +TAP or by right-clicking "Clear All Marker" in the →Measure Bar. Markers can also be named. This name is then displayed in the marker area of the →Time Display when the time line reaches the marker position.

3. The CD-R WRITE Menu

All settings required to burn an audio CD are available in the "CD-R Write" menu. These settings determine the source tracks to be selected (time duration and memory requirements are displayed), the write method (TAO or DAO) and the Finalize mode. After inserting a blank CD-R into the→CD Writer, the CD-R Write menu can be accessed as follows:

VS:	CD-RW/MASTERING → F2 (CD-R Write) ... "Store Current?" → YES
VGA:	CD-RW/MASTERING →CD-R Write ... "Store Current?" → YES

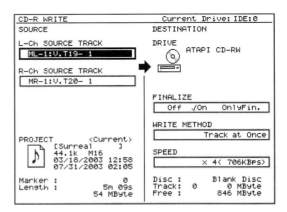

Fig.: 'CD-R WRITE' Menu

CD Writer

SOURCE:
The source tracks for the CD must be entered in the input boxes L-Ch and R-Ch Source Track. As default (and the usual selection) the last recorded Mastering Tracks are pre-selected (V-Track 16 of the highest track number pair). In principle, every V-Track can be chosen as a source track, operating with reasonable care is therefore recommended (e.g. for rehearsals reasons, the piano can be placed on the left and the bass on the right track of the CD).

Tip:
By holding down SHIFT, selection is made in steps of 16 V-Tracks, which is usually faster.

PROJECT:
For a better overview, the number of CD track markers, the playing time and the required storage capacity are displayed. Be careful not to exceed the CD's capacity.

WRITE METHOD:
→Track at Once (p. 226)
TAO can be used to burn one track at a time. In this case, there is no need to set CD track markers. The pauses between tracks on the CD will be two seconds long. TAO does not conform to the →Red Book Standard and cannot be selected for CD-RWs.

→Disc at Once (p. 57)
DAO means an entire CD-R can be burned in one session (automatically finalized). This approach is Red Book-compatible. With the exception of the first CD track, all markers need to be generated manually. As the blanks can be specified freely (e.g. blank= 0), this procedure is ideally suited for medleys and live recordings. A burned CD-R cannot be overwritten.

FINALIZE:
The →Finalize operation copies the information of the temporary table of contents to the final →TOC (Table Of Contents). Finalized CD-Rs can be played back on any commercially available CD player. If necessary, select "OnlyFin" to convert an unfinalized CD-R to a universally playable CD-R.

Track at Once: Select "Off" if there might be other tracks to be added to the CD. "On" must only be selected when burning the last track.
Disk at Once: "On" is selected by default.

SPEED:
The write speed can usually be set to the maximum value. If an error message is displayed, select a lower value.

Press F5 (OK) to check the CD. If the CD-R turns out to be faulty, an error message is displayed. YES starts the write operation.

Disk Image:
The VS can only burn stereo files in the format →CDR to a CD. While CDR format mastering tracks generated in the Mastering Room meet this requirement, regular V-Tracks must be converted to "Image Files" before burning them. This automatic operation requires additional hard disk capacity, which can however, be sourced from any partition. The image file is deleted if "NO" is selected in response to the "Write another Disk?" message that appears after the burning operation.

Copying a CD, LP, etc.
For cassette, CD, LP, etc., i.e copies that don't require additional editing or sound processing, a →CDR-mode Project can be created (→Project New). The stereo track generated during the recording already acts as an image file and can be written to a CD straight away. There is therefore no need to create CDR-format mastering tracks.

CD Writer

A CD-R has a heat-sensitive organic layer which the laser beam can burn bubbles into (so-called "data pits"). These pits and the unchanged areas ("lands") make up the digital information. During playback, the laser detects reflecting and non-reflecting areas that correspond to the values "1" and "0". In some countries, pure CD-Rs and special audio CD-Rs differ in that the latter contain a copyright bit. The VS's burner ignores that bit.

CD Digital Rec

VS-2480
For a VS-2480 without a C., a suitable →SCSI device needs to be connected. Select the SCSI-ID 6 and activate its terminator.

CD Capture

"Capture" (also called "ripping") enables CD tracks to be imported from a disc in the VS's writer to the hard disk—without using an external CD player. If the →CD Digital Rec parameter has not yet been activated, a message regarding copyright infringements is displayed. The transfer speed is determined by the Project's →Record Mode (CDR: 40%, M16/M24: 80% and all R-DAC modes: 150% of the actual track length). Set the →Sample Rate to 44.1kHz. After inserting a CD into the VS's writer, the "CAPTURE" menu is selected as follows:

CD-RW / MASTERING → CD Capture → YES

The following buttons are available for navigating the CD and controlling playback (with the same functions as a regular CD player):

Button	Function
PLAY/STOP	Playback/Stop
ZERO	To the top of the track
NEXT/PREVIOUS	To the next/previous track
F3 EJECT	Eject the CD
F6 (EXIT)	Quit

Table: CD Capture functions

There is also a →CD Player function allowing CDs to be played back.

CD TRACK: The left column lists the CD tracks and their playing times. Move the cursor to the right and use the dial to select a track.

DESTINATION TRACK:
The Destination Tracks of captured material are determined in this field. Move the cursor to DESTINATION TRACK of the left channel ("L-Ch") and select the desired track. Though the destination track for the right channel is selected automatically, its assignment can be changed (+ SHIFT: advance/go back in steps of 16 V-Tracks).

CAPTURE TO:
Specifies the location within the VS track where the imported data starts. Select "to ZERO" (00:00:00:) for the first track. For subsequent tracks, pauses that are 0, 2 or 4s in length can be specified.

OFFSET and LENGTH
These parameters allow the Start and End locations of the CD excerpt to be ripped, to be set:

Track	Offset	Length
Whole	Start of Track	Whole 1 Track
Partially	Specify time	Specify time

F2 (CLEAR) allows to the entered values to be reset.

Tip
- During playback of a CD track, the Start position can be specified in realtime by pressing F1 (→Now). The remaining time indication (LENGTH) is updated automatically.
- For compilation CDs, etc., create a new Project in →CDR record mode (not available on the VS-2000). Doing so significantly shortens the time required for CD ripping and burning of the resulting CD-R.

CD Digital Rec

The "CD DIGITAL REC" parameter has been included as an opportunity to point out copyright issues inherent to digital copies of protected material. Making digital copies of a CD, which involves bypassing the →SCMS copy-protection flag, is only possible if the statements that appear when activating this function are accepted. (VS-2000: SYSTEM menu).

Utility → Global 1 → CD Digital Rec

CD Marker

The VS can also add an SCMS flag to personal recordings (→Digital Copy Protect).

CD Marker

Most CD players allow the following or previous tracks to be skipped or selected by specifying their track numbers. On the VS, CD markers enable this function (→CD Burn). These markers are different from regular →Markers (→2. CD Markers (p. 36)):

Mouse:	Right-click the Measure Bar → CD Marker
VS:	CD/RW + TAP
VGA:	CD/RW + click [TAP]

CD Player

The internal →CD Writer allows the play back of both commercial CDs and unfinalized CD-R/CD-RWs*. Especially during →Live Mixing, this negates the need for an external CD player for music during breaks.

On the VS-2480HD model, an external CD writer needs to be connected to the SCSI port.

CD-RW / Mastering → CD Player → YES

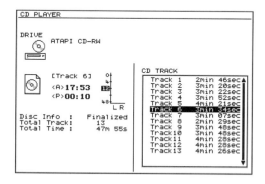

Fig.: CD player inside the VS

In addition to the playing time (P), the display also shows the absolute location (A), which is calculated from the CD's beginning.
To operate the VS's CD PLAYER, the following buttons are available:

Button	Function
PLAY/STOP	Playback/Stop
ZERO	Top of the track
NEXT/PREVIOUS	To the next/previous track
FF/REW	1s FF/REW (by STOP)
F3 (EJECT)	Eject the CD
F6 (EXIT)	Quit

The →CD Capture function enables the recording of entire CD tracks or excerpts to the VS's hard disk.

CD Recover

C. refers to the procedure for transferring archived Backup data from a CD-R back to the VS (archiving can be performed with →CD Backup). This restores the Project's original state, including all audio data on all V-Tracks, all mixer settings, routings, Automix data, etc. (→Total Recall). It is possible to recover several Projects simultaneously from a CD. Larger Projects (>700MB) probably require swapping CD-Rs as the data is loading. Using the →Import function, Projects and Songs created with other VS models can be loaded. The procedure is almost the same.

PROJECT → scroll down → CD-RW (or ATAPI) → List → Mark → Page → Recovr

1. CD Writer As 'Current Drive'

- Call up the →Project List and move the cursor to "CD-RRW" or "ATAPI" drive.
- Press F6 (LIST) + 2x YES to save the current Project and display the CD's contents.
- Select the desired Project with MARK or YES on the CD-R.
- Repeatedly press PAGE until F2 reads "Recovr".
- Press F2 (Recovr)*.
  The "Recover" menu appears.

* To import a Project created on another VS model, press F3 (Import).

2. 'Project Recover' Menu

The "Project Recover" menu specifies the reading speed and the target partition. The "Erase all Projects" parameter should be handled with care.

Scroll:
F2 and F1 scroll the list of Projects on CD.

CD Speed:
In most cases, select the "MAX" setting. If an error message is displayed, select a smaller value.

SelDrv:
Press F4 to select the hard disk →Partition where the CD data should be stored. Partition "0" is selected by default.

Erase All Projects:
Normally, the "Off" setting shouldn't be changed. Select "On" only if all Projects on the target partition may be erased!

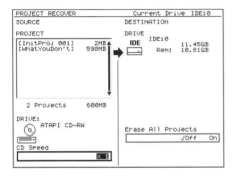

Fig.: Loading an archived Project—Recover

Operation
Press YES and confirm the question that appears, after which the Project data is copied from the CD-R to the selected hard disk partition. Only the hard drive, and not the CD-R, can be used for playing back and editing Projects. Depending on the Project size, the recovery may take a while.

CD-R

The VS can turn a CD-R ("Compact Disc Recordable", "disc") into an →Audio CD (→CD Burn), use it as →CD Backup media or for →Wave Export operations. The available data area can be filled only once, whilst →Finalize allows the disc to be closed. A CD-R has a heat-sensitive organic layer into which the →CD Writer's laser beam can burn bubbles (so-called "data pits"). These pits and the unchanged areas ("lands") make up the digital information. There are several CD-R types with the following storage capacities: 650MB (74 minutes), 700MB (80 min.) and 800MB (90 min.). In some countries, pure CD-Rs and special audio CD-Rs differ in that the latter contain a copyright bit. The VS's burner ignores that bit.

CD-R Write

Use the "CD-R Write" menu to make settings for burning a →CD Audio (→CD Burn).

CD-RW/Mastering → CD-R Write

CD-RW

A "Compact Disc ReWritable" is a CD whose contents can be erased and replaced up to 1000 times. Burning data onto it is similar to the procedure for →CD-Rs. A light-sensitive color layer can be used for the pattern of crystalline (which is reflective) and amorphous (non-reflective areas) states ("phase change technology"). The pattern is obtained through variations of the laser intensity (600°C for digital "1", 300°C for digital "0"). Commercially available CD players usually cannot play back CD-RWs. It is therefore recommended to restrict their usage to VS-specific applications (→CD Backup and →Wave Export).
Before re-writing a CD-RW, it must be erased with the VS (*"Finalized Disk!"* → YES → … *Erase?* → YES).

CDR

The VS's "CDR" setting generates a 16-bit stereo-interleaved →Record Mode, which corresponds to the →Image File necessary for →CD Burning.

CH EDIT

```
ML*1:V.T19- 1
MR*1:V.T20- 1
```

Fig.: Flagging CDR tracks

"CDR" can be applied to all V-Tracks of an entire Project (→New Project) or to specific →Master Tracks in the →Mastering Room. The "*" symbol is used to indicate such tracks. CDR and standard tracks cannot be played back simultaneously. This explains why CDR mastering tracks can only be edited and played back in the Mastering Room. This limitation also applies to "→Phrase New". When trying to mix CDR and standard tracks, the following error message appears:

"Found Illegal Track Pair!"

CH EDIT

A →Channel's "CH EDIT" page provides all tools available for processing the associated audio signal. These include the fader setting, Pan, →Equalizer, →Dynamics, effects and output assignments, etc., which are displayed together in the VGA menu page:

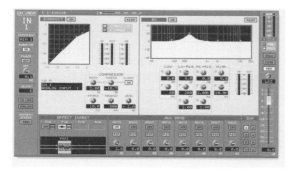

Fig.: VGA's Channel Edit page (Input channel 1)

The "CH EDIT" page corresponds to a physical →Channel Strip of an analog mixing console.

'Input Mixer' or 'Track Mixer'?

Before editing a channel, the →Mixer section to which it belongs must be selected. In most instances, operation takes place in the →Track Mixer channels. The →Input Mixer and →FX Return Channels are also available.

1. VS Operation Panel:
Use the →Fader Buttons to select the desired mixer section. Press the button:
- "TR ..." for Track Mixer
  "IN ..." for Input Mixer)
  VS-2000: CH EDIT
- Press CH EDIT of the desired channel.

2. Mouse In The LCD
Click ⌂ HOME (upper left) and select CH VIEW. Select a channel from the menu.

3. Mouse On The VGA Screen
3.1. Access to all Channel sections:
Menu bar: → MIXER → select a channel from the desired section.

3.2. Track Mixer Only:
Double-click the track number (gray field).

Currently Selected Mixer Channel
The last mixer channel selected remains "current" until another one is selected. Its number appears in the upper left area of all menus, e.g. TR 3, IN 8 or FX2RTN. Using the →Channel Strip enables mixer settings for the currently selected channel to be immediately applied.

VS-2000:
The VS-2000 provides separate buttons for the Input Mixer section (located below the Gain knobs) to call up an Input CH EDIT menu. The VS-2000's Input-Mixer has no Fader, and therefore no fader buttons to assign to a mixer section. Be aware, however, that the TRACK-STATUS buttons have two functions. By default, they are used to set the track →Status. Only once CH EDIT has been activated (button lights) can the Track Mixer's "CH EDIT" page be selected.

CH PARAMETERS

The VS-2000's →Channel Strip allows the desired mixer parameters to be set via the dedicated additional controls. The functions available to be set with the knobs are: →Equalizer or →Compressor/→Expander, →Panorama and a selectable →AUX Send. The CH PARAMETERS encoders first need to be assigned to the desired track channel. To do so, call up the appropriate →CH EDIT page. The name of the selected channel appears in the upper left corner of the display. Input Mixer parameters (→Mixer), on the other hand, can only be set via the display.

EQ (4-BAND EQ):
The gain of the four frequency bands can be set using the knobs. By holding down the EQ button while rotating them, the center frequency can be changed.

DYNAMICS:
As indicated by the knobs, the Threshold Level (Expander: Ratio) and time parameters can be set here.

SENDS:
By default, the Send knob sets the FX1 Send level of the selected parameter. Use the →CH EDIT menu of the desired channel to assign it to another bus. (This can be set for each channel separately!) Click a check box to select the output destination for the SENDS.

RSS/PAN:
In most instances, this knob controls the selected channel's Pan. When activating →RSS PAN, the channel signal can be placed anywhere in a 3D circle around the listener. This effect, however, requires the processing power of an entire effects board.

Channel
Channels are the fundamental parts of any audio mixer. They are usually fitted with a fader, an →EQ, →Dynamics, →Effects, etc. The →AUX Busses and →Direct Path functions allow signal "copies" to be sent to other destinations within or outside the console (on the VS, that would be the internal effects or an external headphone amp). In addition to the comprehensive Input Mixer and Track Mixer channels, there are slightly more basic channels such as →FX Return Channels (no dynamics, no EQ)

and input channels 9~16 on the VS-2400. The →Fader is probably the most important tool of a channel. It sets the channel's level in the →Mix Bus. All controls available for a channel are collectively referred to as "→Channel Strip". Be aware that the channels are not the same as the →Tracks (audio data on hard disk).

In the VS, the channel tools are to be found in the →CH EDIT menu. The VGA screen can display several channels simultaneously. Use PAGE 2/3 + F2~F4 to select them. The VS models provide the following number of mixer channels:

Model	Input-Mixer	Track-Mixer
VS-2480	24	24
VS-2400	16	24
VS-2000	10	18

Note:
There are more channels in the Input-Mixer than physical inputs available. The connection takes place in the EZ Routing's →Patchbay.

Channel Link
Combining two neighboring mono channels to a stereo pair is called a "Link". Linked channels are always edited and processed in tandem. This often proves to be a major time saver. Here's how to establish/break a Link:

VS Operation Panel:	CH EDIT button + button of the adjacent channel
CH EDIT:	Activate CH LINK
VGA PAGE 2/3 (F1~F4):	Open the LINK flip menu below the faders.

Fader & Pan
In linked channels the Fader and Pan control changes in their display of icon. The VS-24xx's motorized faders move in tandem, either fader or →Panorama parameter can be used to change the corresponding setting of both channels. On the VS-2000 (which doesn't have

Channel Strip

motorized faders), the fader of the left stereo track is active. Level or Pan corrections for just one channel of a linked pair can be applied using →Offset Level in the sub display. This allows the stereo image to be balanced and/or to narrow the image down

Sub Display: In the LCD's →CH EDIT menu, move the cursor to "FADER" or "PAN" and press YES. On the VGA screen, click "SUB DISP".

LCD's Parameter View: The LCD enables at least four linked channels to be displayed simultaneously via →Parameter View. To display this menu press F6 (PrmV) after moving the cursor to the "FADER" or "PAN" fields.

Automix:
For the faders and Pan parameters, contrary to normal channels, linked channels generate special →Automix data called "OFFSET LEVEL" and "OFFSET PAN". Choose these offset parameters for graphic and micro editing.

Fader Link
There is a special kind of CH LINK called "→Fader Link". It only links the faders of a channel pair.

VS-2000:
Track channels 15/16 and 17/18 are already stereo pairs (and cannot be separated).

Channel Strip

Due to cost saving measures, most digital mixers only provide physical faders, while all other settings have to be adjusted via the display. There is usually one additional set of channel strip controls that provides direct access to all parameters available for the various channels. This strip can be assigned to the channel about to be edited. (The VS-2400 has no C.)

Selecting A Channel
Always start by selecting the mixer channel to be edited, which is automatically assigned to the channel strip controls. The name in the upper left corner (→CH EDIT) refers to the currently selected channel.

VS-2480:
By activating →PRM EDIT, the "PAN" knobs (encoders) are available to be used as a channel strip* with the following functions: knobs 1~5 adjust Dynamics parameters, while knobs 6~16 are assigned to the EQ. See also the labels printed on the front panel. Additionally, the "→Prm Knob Auto Display" allows a special mode of operation so that the corresponding LCD page of the parameter which is currently set with the knobs appears automatically.

> * *The knobs can also be used as AUX Send controls of 16 channels at a time (→AUX Send). That function, however, is not a channel strip.*

VS-2000:
The →CH PARAMETERS section allows the →Equalizer or →Compressor/→Expander, →Panorama and a selectable →AUX Send bus to be assigned to the knobs of this limited channel strip.

VE-7000
An optional →VE-7000 controller can also be used as an external channel strip. In addition to the knobs discussed above, it also provides a joystick for →Surround Pan settings.

Channel View

The LCD provides two modes for the control of viewing and editing →Channel Strip parameters: CHANNEL VIEW and PARAMETER VIEW. When calling up the →CH EDIT menu, the VS is principally in C. This mode allows several parameters of the selected mixer channel to be accessed on one page. By contrast, "→Parameter View" displays only one parameter for several mixer channels. Use the F6 button to select the desired mode.

CH Edit → F6 → PRM.V / CH.V

If an external monitor is also connected, the "Channel View" page can be displayed at all times in the LCD (→ID ChV). The VGA provides a "→Multi Channel View" with a great number of edit parameters available.

Chorus

A chorus effect splits the incoming signal into two lines and delays one line by up to 30ms, which is then modulated. In combination with the adjustable frequency (speed) and modulation depth, the rather significant delay produces a broad, hovering sound. Generally chorus effects have a monaural input and stereo outputs, that's why this effect is used to make a signal appear larger than it actually is. The VS provides the following chorus algorithms in the →Effect Patch: "Space Chorus", "Chorus RSS", and "Stereo Delay Chorus". They all have →True Stereo inputs. The →Vocoder, →Vocal Multi and →AnalogFlnger ("Mode"= Cho) also provide a chorus block.

Chorus

Effect patch:	P091~096, P009
Connection:	Send/Return
Application:	Standard

Chorus RSS
This algorithm adds an →RSS unit to the →Chorus effect, which creates a unique 3D (three-dimensional) depth. At low Send levels, "Chorus RSS" is ideal for acoustic guitars and for broadening signals.

Stereo Delay Chorus
This patch provides three blocks: →Delay, Chorus, and EQ. Thanks to the Delay block's Feedback parameter, it can add echoes to the effect sound and can be set in such a way that the chorus effect becomes even larger than it already was.

Space Chorus
This emulation of Roland's legendary SDD-320 "Dimension D" is extremely easy to set. "Space Mode" allows one of the original unit's possible button combinations to be selected, and also the resulting effect intensity.

ChromToneAmp
→McDSP ChromeToneAmp (p. 119)

Cinch
C. (RCA) is an →Unbalanced connector type usually found on consumer devices (analog mono signals). It is also used for digital →S/PDIF signals (digital L/R).

Clean Up
To delete unnecessary audio portions, the VS offers the editing functions
→Region Erase (p. 190)
→Phrase Trim (p. 157)
→Phrase Delete (p. 149).

CLEAR
The CLEAR function is used to remove Markers, Scenes, Locators, Punch In/Out positions, loop points, etc.:

CLEAR button: Hold down C. on the VS and press the button of the parameter that is to be cleared. Alternatively click the parameter icon in the →Measure Display (while holding down C.)

Function button: Within the Marker, Locator, Scene, Loop, and Punch In/Out menus, using the F2 button clears the assigned values.

Mouse: Right-click the →Measure Bar to delete all Marker positions ("Clr All Marker"). Alternatively, hold down the CLEAR button and click a Marker, Loop, Punch In/Out or Locator position.

0dB Position:
Holding down C. while pressing a CH EDIT button recalls the "0dB" level setting for the corresponding mixer channel.

Bypassing quantization:
When the →Grid is active, holding down C. while editing with the mouse temporarily ignores the quantize setting.

Clipping
While overdriving an analog input circuit leads to slight (often desirable) or serious distortion, overloading an →A/D Converter causes all bits to be set, producing a harsh distortion, which is called "clipping". This is why →Level Setting

While Recording is of the utmost importance on digital audio systems. For this, it is imperative to maintain the →Headroom.

Clock
A C. signal (digital sync, wordclock) electrically displays a square wave whose frequency corresponds to the selected →Sample Rate or a multiple of it. It ensures that the sample words of the transmitting and receiving digital audio devices are in perfect sync. Deviations from the clock lead to clicks, pops and other audible degradations in sound. Note that sync problems may also be due to the use of the wrong cable type (→Digital Cable). Digital clock signals can be transmitted or received via dedicated →Word Clock cables (VS-2480 has an wordclock IN). The self clocking formats →S/PDIF and →AES/EBU are much easier to handle because the clock is embedded in the digital audio data stream. A limited number of S/PDIF and AES/EBU devices can then be daisy chained. For→Recording of digital Signals with the VS, the appropriate →Master Clock setting must be selected.

Clock and Midi Synchronization
The VS can be simultaneously synchronized with →MTC timecode (both recorders play at the same speed) and wordclock (both recorders play at the same sample rate), the resulting synchronization signal is of an entirely different nature: possible tempo fluctuations have no effect on the pitch and duration of digital audio signals because those aspects are controlled by the internal quartz oscillator's sample rate. In other words: clock and MTC must act together. If digital audio signals are recorded from a unit and the unit is also used as a master in the MTC synchronization "External Timecode" must be selected for the parameter →Master Clock.

ClrPrt
The "Clear Partition" parameter in the →Project List enables all projects from the selected →Partition to be deleted.

Cluster
A C. is the smallest data unit on a →Hard Disk that the VS can read. The number of faulty C.s can be checked with →Drive Check.

Coaxial
The digital →S/PDIF audio format can be transmitted as electrical ("Coaxial") or light signals ("Optical"). The VS-24xx models provide both connector types for the input and output sections. The VS-2000 only has coaxial digital sockets.

Comb Filter Effect
The combination of two identical signals that are slightly offset with respect to one another can lead to cyclical gain boosts and cancellations that produce a nasal, metallic sound. If the offset is a multiple of half the wave length (e.g. 180°), the two signals cancel each other out. This is, however, only the case with odd-numbered harmonics such as 3, 5, 7, etc. Even-numbered harmonics are amplified.

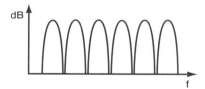

Fig.: Comb filter effect

This physical phenomenon is useful to the→Flanger, →Phaser, and →Chorus →Effects. It may also occur when several microphones are used to capture the same signal.

Combine
→Project Combine (p. 166)

Comment
In addition to naming a Project, the comment field leaves space for information such as musician's names, settings of external effect processors etc. There are 100 ASCI characters free for this feature (→Name Entry). The LCD's →Screen Saver can display such comments.

Compact Studio

A compact studio provides all sections needed for studio productions in one unit, more specifically:

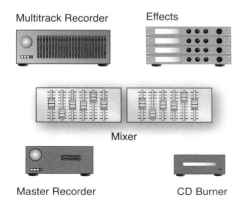

Fig.: Components of a compact studio

- Analog and digital inputs
  (→Microphone Preamp (p. 122),
  →Phantom Power (p. 145)).
- Recording mixer (→Input Mixer (p. 97))
- →Patchbay (p. 144)
- Digital Multitrack Recorder
  →Hard Disk Recorder (p. 89)
- →Track Mixer (p. 228)
- Headphone amplifier, →PHONES (p. 146)
- Independent effect processors
  (→Effect (p. 62))
- Digital editing functions (→Editing (p. 62))
- Master recorder, →Master Tracks (p. 118)
- Analog and digital outputs for mixing, and additional →Outputs (p. 139)
- →CD Writer (p. 38)
- →Synchronizer (p. 218)

The major advantage of such a system is that all components are already wired up, while all sections can be programmed simultaneously (→Scenes). Furthermore, a DAW is extremely compact (→Live Recording), and its function can often be expanded (→System Update) etc. The only drawback is the system's complexity and the steep learning curve for users.

Compressor

A C. reduces the →Dynamic of processed audio signals by attenuating signal peaks. Its operation can be emulated by fader movements, which is still being done for classical recordings. A C., however, has a faster response time and is usually more accurate than a manually working engineer. "Look-ahead" compressors, for instance, use a slight delay that gives them time to prepare their response to upcoming signal peaks (→Mastering Tool Kit).

Threshold and Ratio

In most instances, a compressor is expected to only attenuate loud passages (i.e. levels that exceed a certain value). The level at which a compressor kicks in can therefore be set with a parameter called "Threshold". This specifies the upper limit of signal levels that are passed through without being affected. Signal levels that lie above that threshold are reduced. The "Ratio" parameter specifies exactly by how much excessive levels are to be reduced. Low Ratio values produce slight level changes, while high values result in significant attenuation. The Ratio parameter works proportionally: At a "1 : 2" ratio setting, signals levels 10dB above the Threshold level are halved (divided by 2) and thus reduced by 5dB. This also means that the dynamic range is reduced by 5dB.

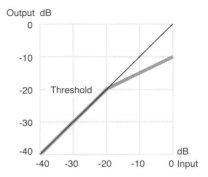

Fig.: Compressor Threshold

Compressor

Signal levels below the Threshold are "processed" with a Ratio of "1 : 1", meaning that they don't change. It follows that a compressor reduces the overall level. The amount of gain reduction is displayed by the "GR" level meter. If necessary, the loss can be compensated for by increasing the compressor's output level.

A special variety of the compressor family is called a "limiter". Ratio values in excess of "1 : 20" no longer reduce level peaks proportionally but simply flatten them. The following illustration "Compression and limitation" shows the level of a signal that is processed by a compressor and a limiter. After the dynamic processing the Compressor/Limiter's output level is set to the highest possible value. The dynamic range's reduction to a few dB is clearly visible.

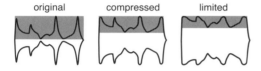

Fig.: Compression and limitation

Attack And Release

With "Attack", the start of compression can be delayed even though the signal exceeds the Threshold. In contrast, "Release" enables the compression to be maintained even though the levels fell below the Threshold. Longer Attack times will emphasize the onset of subsequent signal pulses, while longer Release times result in compressing signals whose levels lie below the Threshold level.

VS's Channel Compressors

The VS provides separate compressors for all mixer channels (→CH EDIT). Their default settings have been chosen in such a way that they can be used for almost any signal (use ON to activate). For a stronger compression raise the Ratio value. To only flatten out signal peaks ("limiting"), look at the IN level meter to see at what level the peaks occur. Next, set the Threshold level to a few dB below that value and select a rather high Ratio value. The narrow setting range of the VS's Threshold parameters may sometimes prove a serious limitation. Compression is next to impossible for signals below −24dB.

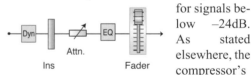

As stated elsewhere, the compressor's input level cannot be controlled using the attenuators and inserts. Therefore, the →Normalize and/or →Phrase Level functions can be used to increase the level of the recorded phrase. (Alternatively, route the signal to an output, and from there back to an →Input Mixer channel and process the signal here). As with many aspects, setting the compressor parameters should first and foremost be a matter of listening to the result rather than entering parameter values seen elsewhere, or following preconceived ideas. For powerful compression, use the →Mastering Tool Kit algorithm as →Insert effect for the channel needing to be processed. Yet another possibility is to add the →Urei 1176 Compressor and →LA-2A →Plug-ins to the VS setup for classic studio compression.

AutoGain:

Compressors almost invariably reduce the level of the signals they process. To achieve the wanted increase of loudness, the VS provides an "AutoGain" parameter that makes up for such losses. This explains why a channel tends to become significantly louder when its compressor is switched on (AutoGain is active by default). To avoid clipping that might result from automatic gain corrections, the compressor is connected to a →Limiter. The limiter's functions are hidden and cannot be adjusted. Another advantage of maximizing channel levels is that converting 24-bit signals to the dynamic range of a CD (16 bits) tends to produce better results (because the available bits are used more effectively). Be aware that the limiter may produce sound coloration. It is therefore recommend switching "A.Gain" off for percussive signals (or at least trying to do without).

Key In
In most instances, the same signal being processed is analyzed to establish the required gain reduction. With "Key In" the compressor can use another channel's signal for its analyses. This special function is called "Ducker" and it allows typical voice-overs with automatically lowered background music (→Key In).

Condenser Microphone
Thanks to their natural and transparent transducing qualities (and due to their mechanical delicacy), condenser microphones are usually only used in recording studios, mainly for vocals, acoustic instruments and ambience. The C.'s diaphragm (a piece of 1~10μm foil) acts as an electrode that is located at 5~50μm from another electrode, the two of which constitute a condenser. The electrode's capacity changes in response to the movements caused by the sound waves. Based on a constant DC voltage that is applied, the plate movements (and changes in capacity) can be translated into proportional voltage fluctuations. The amplifier built into the microphone boosts the ultra weak alternating current to →Microphone Level and takes care of the impedance conversion. The polarizing voltage and the power to the preamplifier are usually supplied by →Phantom Power.

Contrast
If the LCD proves difficult to read, its contrast can be changed with the CONTRAST knob.

Control Changes
C. are standardized MIDI messages that can be used to control certain parameters of MIDI sound generators, mixers, effect processors and →DAWs. In the VS →MIDI Mixer Settings using either CCs or →System Exclusive Data can be transmitted or received.

Control Local Sw
Via C., the VS's faders can be disconnected: This is usually only necessary to avoid MIDI loops when the VS transmits settings to →MIDI Sequencer while also receiving data from it. In doing this, access to the VS can be refused for unauthorized users.

Utility → MIDI Parameter → "Control Local Sw"

Control Surface
A C. is a unit with faders, knobs and buttons that allows the control of MIDI or audio devices and software. The VS-2480's faders and PAN knobs can be used for remotely controlling software like "Logic", "Cubase", "Pro-Tools" and "Digital Performer". To do so, the appropriate driver needs to be downloaded from *www.rolandus.com*.

Copy
The VS provides the following functions for copying audio and Automix data to a desired position:
- Editing Regions (between the IN and OUT points, which are defined by drawing a rectangle with the mouse or entering the appropriate values):
→Region Copy (p. 186)
- Editing Phrases (entire "bars") of recorded or divided portions, that have been separated from the original Phrase):
→Phrase Copy (p. 147)
- Editing all tracks of the Playlist in various areas, built by Markers:
→Region Arrange (p. 184)
- Editing Automix events:
→Automix (p. 19)
- VS-2480: Editing recorded Phrase Sequencer events:
→Phrase Sequence (p. 155).

Copy Mixer Parameter
The parameter settings of a mixer channel can be copied to other channels. It is even possible to specify whether to copy only the →Dynamics or →Equalizer settings, or all parameters. Both Input Mixer and Track Mixer can be selected as the copy source.

CH Edit → Page 1/5 → F5 (CpyPRM)

Copy Mixer/Scene Parameter

Source Ch
The settings of this channel serve as a source for the parameter copy. Any →Input Mixer or →Track Mixer channel can act as source.

Copy Target
"All" means that all parameter settings are copied. "Dynamics" or "EQ" mean that only these settings will be copied from the source track.

Copy Mixer/Scene Parameter

When creating a new Project (→New Project), C. allow the current mixer and Scene settings to be copied to a Project about to be created. To include all settings of a given Project in a new one, use →Project Copy.

Tip:
With the help of an external midi sequencer, VS Scenes can be stored and recalled into the current Project at a later stage (→MIDI Bulk Dump).

Copy Utility Parameter

When creating a →New Project, all general settings may be adopted (→UTILTY Menu**).** This allows existing synchronization, MIDI, Play/Rec and Tempo Map settings to be reused. The Generator/Oscillator and Automix parameters are not copied by this function.

COSM

"Composite Object Sound Modeling" is a technology developed by Roland. It is used in the →Effects algorithms of the →Microphone Modeling, →GuitarMulti and →Speaker Modeling groups to simulate physical phenomena.

CpyPRM

→Copy Mixer Parameter (p. 49)

Crossfade

Crossfades are generated automatically by the VS when editing audio data or recording in Punch In/Out mode. This eliminates unpleasant click and pop noises. "Fade Length" allows the duration of such "superimpositions" to be specified (2~50ms). "Fade Curve" is used to select the shape (exponential or linear) of the crossfades. Note that this function does not correct defects that may arise due to the use of inappropriate editing locations.

Utility → PlayRrec 1 → Fade Curve / Length

Current Project

C. refers to the currently loaded Project, which can be played back, edited and saved e.g. *"Save Current Project?"*.

Current Time

The current time location is indicated by the red →Position Line (graphically) and shown in the →Time Display (numerically). In this way, it is possible to read the value of the current time location within the project. Dragging the Position Line or changing the Time display's value makes it possible to →Navigate within the project. →Locators, →Markers, →Edit Points and →Automix Snapshots are always generated at the C.

Cursor

The mouse cursor allows the selection of a position within the →Operation Display. It can be used both on a VGA screen and in the LCD. The cursor icon changes according to its location (e.g. above certain objects) and status, which is a great help for →Editing operations.

- ▸ Location and display
- 🖑 Click for "on/off" parameters
- ↕ Click and drag for vertical input
- ↔ Click and drag for horizontal input
- 🖑 Passages can be moved (see below)
- ✊ Hold down the button to move
- ✊ Hold down SHIFT and move to copy
- ⇥[To trim the beginning of the Phrase
-]⇤ To trim the end of the Phrase

By holding down the left mouse button and dragging the mouse over a parameter, its value changes. Within the input menu and →CH EDIT environment, the C. can also be moved using the →Cursor Buttons.

Cursor Buttons

These buttons allow the cursor to be moved without using the mouse. On the →HOME page, ≑ enables track selection (→Track Select), while ◆▶ can be used to move to the desired time, measure or Marker field of the →Time Display. While holding SHIFT, the C. can be used to zoom the →Playlist and the →WAVE DISP.

Cut

The Cut →Editing function allows unwanted data to be removed from a specified region. Subsequent passages are then shifted to the left by the same amount. This operation is similar to splicing a tape and removing a section. The VS provides the following C. functions: →Region Cut (p. 188) and →Automix (p. 19). In contrast, →Erase discards unwanted passages without moving passages.

Cut (EQ)

Reducing the level of an →Equalizer frequency band (→Gain) is often referred to as a "cut". (Increasing the level is called "boosting").

D

D.I. Box

A "direct-injection box" reduces the impedance of high-Z signals (guitar, bass), making them usable for low-impedance (analog) pre amplifiers. It suppresses ground loops and interferences by converting →Unbalanced signals to →Balanced ones and establishing a galvanic separation between both devices. Except in rare cases where faced with severe →Hum problems, a D. won't be necessary for the VS. Normally the →Guitar Hi-Z input is perfect for a loss-less connection of electric guitars and bass guitars.

D/A Conversion

D. transforms →Digital Audio Signals back into analog signals. Similarly to the →A/D Conversion, the →Word Length governs the dynamic range, while the converter's linear quality controls the degree of harmonic distortion. The audio signal coming out of the converter still exhibits clear-cut steps and is therefore processed by a steep low-pass filter.

DAO

→Disc at Once (p. 57)

Data Dial

Parameter and time values can be entered with the D. (TIME/VALUE). From a technical point of view, it is an encoder (alpha dial).

1. Locating A Position Within A Project

On the →HOME page, the TIME/VALUE dial allows the →Timeline to be moved within the current Project. The size of the incremental/decremental steps is governed by the cursor's position in the →Time Display.

2. Entering Values

Whenever a parameter is selected, its current value is displayed in reverse and can be modified with the dial.

Changing the Step Size

Hold down SHIFT while turning the dial to increase/decrease the step size by a factor of 10.

Date/Remain Sw

D. allows to display in the LCD either the current date or the remaining recording time based on the available hard disk capacity.

Utility → Global 2* → Date/Remain Sw

With →Remain Display Type whether the remaining hard disk capacity should be displayed as a time, MB or "%" value can be specified. Alternatively the number of remaining →Events can be displayed.

* VS-2000: System Param2 rather than "Global"

Date/Time

The date and time set on the VS will be shown in the display (optional). D. also automatically time-stamps →Takes and →Undo levels. In this way, operations can be located based on their dates of creation or modification.

Utility → Date/Time

To change the setting, move the cursor to an entry in the "DATE" or "TIME" field and use the VALUE dial. Alternatively, click an entry, hold down the mouse button and drag the mouse. "Date Format" specifies the order in which the year, month and day are displayed. Normally, the setting is "dd/mm/yyyy". The calendar, however, only displays two months at a time.

While the menu bar of the VGA screen always displays the date and time, the LCD can only display either the remaining storage capacity or the date and time (→Date/Remain Sw).

DAW

A "Digital Audio Workstation" is a comprehensive unit comprised of a digital mixer, a hard disk recorder, effects, mastering and CD functions. In this way, a VS can also be considered a DAW or →Compact Studio.

dB

Decibels are a logarithmic unit for measuring the relative relationship of electro-acoustic values. Even though the level value range is surprisingly wide, the dB unit allows the range to be narrowed to 0~140dB. The ratio "1/2" is described as 6dB, while "1/1000" corresponds to 60dB. A level of "0dB", on the other hand, means that the signal level is neither raised, nor lowered – a fader or trim knob therefore has no influence on the signal. At a 0dB setting, the "pre-" and "post-fader" levels are identical. By contrast, the absolute voltage levels, "dBu" and "dBV", represent physical entities. European countries use a reference voltage of 0dBu= 0.775V, which corresponds to the studio-quality level of +6dB and a voltage of 1.55V. Line-level (–10dBV) therefore corresponds to 0.316V. Another common reference is 1V= 0dB. Sound pressure is another physical unit called "dB_{SPL}" (Sound Pressure Level). "$0dB_{SPL}$" refers to the hearing threshold level at a sound pressure of 2.1×10^{-7}Pa (Pascal). 100 $0dB_{SPL}$ corresponds to 2Pa, and 140 $0dB_{SPL}$ to 200Pa.

dB	Ratio	dB	Ratio
0	1	40	100
3	1.4	60	1000
6	2	80	10,000
12	4	100	100,000
20	10	120	1,000,000

De-Esser

A D. is used to attenuate high-level fricatives and similar high-frequency signals. Similar to a →Compressor, it only processes signal levels at and above the selected cutoff frequency. This means that two conditions must be met in order for it to process:

1. The signal needs to contain frequency portions that lie above the selected FREQ value.
2. The level of those frequencies needs to exceed the threshold (selected with SENS).

The amount of reduction can be set to individual taste. Too much De-Essing produces a lisping effect in the human voice. The following →Effect Patches (multi-effects) provide a De-Esser block: →Vocal Multi (p. 238), →Vocal Channel Strip (p. 238).

Defragmentation

D. refers to the act of combining data that belongs together, but resides in different areas of the active →Partition partition (→Fragmentation), into a single continuous block. The VS doesn't have a D. function. If fragmentation slows down hard disk access, it is necessary to →Format the entire hard disk. Before even considering this measure, →CD Backups should be made and →Plug-ins (if any) uninstalled.

Delay

Second only to →Reverberation, a delay unit is an extremely popular →Effect and hence frequently used during a →Mix. It generates single or multiple echoes of a signal and repeats these according to the "Delay Time" parameter and positions them in the center or L/R. Delays are often used as →Send/Return Effects, because the dry/wet balance can then be set using the Send parameter.

Delay

Effect patch:	P078~P090
Connection:	Send/Return
Application:	Standard

The VS contains the following →Effect Patches that are exclusively used for delay functions: P078~P086. P087~P090 contain emulations of Roland's 201 analog tape echo unit. As an additional block with fewer adjustable parameters, a D. will also be used in Multi-effects such as →GuitarMulti, →Vocal Multi, etc.

Editing

The most important aspect, the delay time, can be set with *Time* (in ms). Most presets are configured as stereo delays, which explains why the following parameters can be set for both the left and right channels. *FeedBackLvl* sets the number of repetitions. This is based on a technique that inserts the output signal once again into the effect's input. (Extreme values may cause feedback. Negative values, on the other hand, invert the phase.) The intensity of the delay can be set with L/R FX Level.

Mono Delay

Most presets are configured as stereo delays, which means that the echoes alternate between the left and right channels. Here, the *Shift* parameter is used to offset the two delay lines. If a mono delay is needed, (whose repetitions occur at the center of the stereo image), it is necessary to set this parameter to "0".

Tip

Short delay times can also be generated with the →MicModeling algorithm when it's used as an insert effect. Here, the delay time is specified in cm.

How To Calculate Delay Times

The following table explains how to ensure that the delay effect runs in sync with the song tempo (without actually syncing it.)

$$\quarternote = \frac{120.000}{Tempo} \qquad \overset{3}{\eighthnote} = \frac{40.000}{Tempo}$$

$$\halfnote = \frac{60.000}{Tempo} \qquad \eighthnote = \frac{30.000}{Tempo}$$

Table: BPM calculations

Divide the indicated value by the song tempo (BPM) to find out which value to set. The value refers to milliseconds.

Example: Tempo ♩ = 120:
Quarter note= 60.000/120= 500 ms
Half triplet= 40.000/120= 333 ms

The "Tempo Mapping Effect" →Plug-in supplied with each →VS8F-3, automatically adapts the delay time to the Project tempo.

Delete

The "Delete" edit function disposes of entire Phrases (→Phrase Delete). In the VS-2480 it also can delete→Phrase Sequence events. As with the →Erase Region Edit function, subsequent passages are not shifted towards the beginning of the song.

Delta Sigma Converter

Unlike the more common variety, a converter of this type (also called "bit stream" or "1-bit" converter) doesn't quantize the absolute level of samples, but rather the level differences between two adjacent samples (higher than for the previous sample: yes/no). Thanks to the 64-times (or higher) oversampling technology, the

Depth

number of samples taken increases, which, in turn, produces a more faithful digital rendition of the analog signal.

Depth

The D. parameter of an →Effect specifies how strongly the incoming signal is processed or modulated. The "Rate" parameter, on the other hand, sets the speed.

Destination

D. refers to the target area or place for the selected operation (→Bouncing, →Track Import, →Region Edit, →Phrase Edit). It follows that the "→Source" is where the data comes from.

Destination Drive

Operations like →Project Copy, →Project Recover, etc., require the specification of a destination drive. This can be performed using the "SelDrv" parameter. It usually refers to a →Partition (IDE:0~IDE:x).

Device ID

When needing to transfer →System Exclusive Data to several VS units of the same model, the Device ID number (1~32) enables targeting a specific unit. When working with only one VS, the D. is of no consequence. Particularly with the →MIDI Bulk Dump function, it is important to look out for identical IDs. Otherwise, the unit will ignore the data. The D. is stored in a →Scene.

DIF/AT

This discontinued 20-bit interface allows the connection of devices in ADAT- and T-DIF-format to the R-Bus format.

DIF-AT24

This 1/3rd rack-sized unit converts →R-Bus signals into the →ADAT format and vice versa. It supports format up to 24 bits/48kHz and allows the connection of a VS-24xx to TosLink equipped audio cards and/or digital mixing consoles.

Fig.: ADAT interface via R-Bus

The D. also provides two additional MIDI sockets for →Synchronization purposes via MTC. Powered via the R-Bus link to the connected VS, it doesn't require an external power supply.

Application
Transmitting/receiving eight channels of digital audio to/from any device with an ADAT interface (e.g. ADAT recorders, etc.).

Digital

Unlike analog devices, digital units no longer use values that change in accordance with the original sound pressure fluctuations. Instead, signals are translated into a series of binary numbers that can be saved, copied, edited and transmitted loss-free. The most important stage at the beginning of each digital recording is the conversion of analog voltages to the corresponding figures in order to encode them as digital information. This is called→A/D Conversion. The →Sample Rate governs the frequency range available to the digital signals, while the →Word Length controls the dynamic range. As they are only series of numbers, →Digital Audio Signals can be recorded and stored loss-free. Through mathematical algorithms, without any additional hardware, they enable changes of various aspects (time compression, pitch, speed, sound, etc.). In order to be heard again, digital data needs to be converted back to analog signals (→D/A Conversion).

Digital Audio Signal

While analog signals consist of constantly changing voltages, →Digital audio signals are, in fact only binary numbers (i.e. codes). This transformation is called →A/D Conversion. Without any decrease in quality, D. can be copied, transmitted and stored seeing as it is only conventional computer data. This data can be processed using various mathematical algorithms. Digital signals, impossible to hear, require a →D/A Conversion to the analog domain before they become audible again.

Digital Cable
→Digital audio signals should be transferred via special coaxial cables to avoid data loss that may occur during the high-frequency (±3MHz at 48kHz) analog transmission of the audio signals that carry them. Cables tend to filter the high-frequency content, because they constitute a condenser and a capacitor. This leads to various frequency ranges being attenuated in different ways, which, in turn, leads to delays (wave resistance). High frequencies are attenuated more strongly, so that the audio information (several MHz) is late with respect to the →Word Clock signal (usually 44.1kHz). This causes phase shifts between the audio and clock signals, which may lead to the clock already referring to the next sample, while audio data for the previous one are still being received. Such shifts produce audible clicks. The following table shows the recommended cable impedance for the various digital signal formats:

Digital format	Cable	Resistance	Terminal
S/PDIF coaxial	Unbalanced	75Ω	RCA
S/PDIF optical	Optical	-	Optical
AES/EBU	Balanced	110Ω	XLR
R-Bus	R-Bus cable	110Ω	R-Bus

Note:
When working at the 96kHz sample rate, it is recommended to use R-Bus cables with a maximum length of 1.5m.

Digital Copy Protect
D. (the copy-prohibit bit, →SCMS) can be activated to avoid unwanted copies of ones's own songs:

Utility → Project Parameter* → Digital Copy Protect

VS-2000: "Digital" rather than "Project Pm".

Digital I/O Parameter
This menu, only available on the VS-2000, allows the following digital parameters to be set:
- Digital master/slave synchronization: →Master Clock (p. 111)
- Converting to lower bit rates (downsampling): →Dither (p. 58)
- Activating SCMS: →Digital Copy Protect (p. 55)
- Selecting the digital input: →Digital In Select.

Digital In Select
The D. on the VS-2400 allows the S/PDIF input (optical/coaxial) to be preselected. After doing this, the "→Clock" setting can then be changed from "Internal" to "Digital In" (see Recording of digital Signals).

Utility → Project Parameter → "Digital In Selct"

Digital Inputs and Outputs
The VS provides special sockets for receiving and transmitting digital audio signals. They support the following formats: →S/PDIF and →R-Bus. With an external (→AE-7000) interface, even →AES/EBU-format signals can be received and transmitted (not supported by the VS-2000). Connections, synchronization, and routing are to be found in →Recording of digital Signals (p. 181)

Digital Recording
→Recording of digital Signals (p. 181)

Direct Outs

Direct Outs

The VS provides a D. function allowing individual track signals to be sent to a PC or any other recorder. Track signals can be routed to the →Analog Outputs or to the →R-Bus. To do so, select the "OUTPUT" page of the →EZ Routing environment and activate →Track Direct Out (not available on the VS-2000).

Tip VS-2000:
Despite the absence of →Track Direct Outs, →Direct Paths can be assigned to up to eight tracks in the EZ Routing's OUTPUT field. Start by routing the physical outputs to the DIR busses (for example):
MST → DIR1/2, AUX → DIR3/4,
MON → DIR5/6, PHONES → DIR7/8.
On the required track channels, use DIR to feed the →Direct Paths.

Direct Path

A D. is able to send *one* signal to another destination inside the VS, or to an output socket. Similarly to an →AUX Bus, however, it only transmits a *single* signal at a time. AUX busses can also be used for Direct Path applications. It is nevertheless recommend to keep the AUX busses for other, more important, purposes.

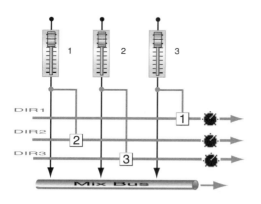

Fig.: Direct Path

Source:
Use →CH EDIT to select a channel strip. The DIR 1~8 switches let the channel's signal to be fed into the desired Direct Path.

The VGA's "→MltChV" page and the LCD's →PRM.V page also provide quick access for path routings. Click the desired switch with the mouse, or use the cursor buttons to select and the dial to activate. A Direct Path can be assigned to only one channel. By selecting a path that's already used by another channel, the previous assignment is cancelled.

Example:
Direct Path 5 is currently assigned to Input Mixer channel 7. If Path 5 is then assigned to Track Mixer channel 15 (by accident), the assignment to INPUT channel 7 is canceled.
Generally a Direct Path can be assigned to any channel—even FX Return channels.

Pre/Post
In The →Master Section's "VIEW" page (see the illustration below), specify whether the Direct Path signals should be sourced before or after the faders of the assigned channel. It can also be used to check the current Direct Path assignments. (But they cannot be edited here.)

Parameters
As shown in the illustration below, there are Pre/Post switches and dedicated controls for setting the Direct Path levels. The path routings can be checked but not edited here.

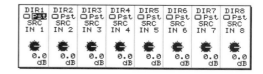

Fig.: Direct Path parameters (Master section)

The controls can be found on the →Master Section's "VIEW" page. Normally, the default setting of "0dB" is satisfactory.

Destination: VS output sockets
Like the AUX busses, the Direct Paths can be routed to the physical outputs on the "EZ Routing" page (→Output Assign). This function can be used for the following:

- Click track for drummers.
- Efficient control of *one* external effects processor, which can be used both by a Send/Return and an insert effect (→Effect).
- Routing a stereo track to a digital output.
- Listening to single signals via headphones.
- Assigning an external Tuner to a desired source.

Destination: within the VS
Internal routings of the Direct Paths can be set with →Routings.

- Assigning an effect to one channel only, →Loop Effect Assign (not available on the VS-2000).
- Switching signal sources while recording to a track (see below).

Recording 'Direct Path' Signals
Besides outputting a Direct Path signal, it can also be recorded (→Bouncing). Use the "EZ Routing" environment to assign the desired DIRECT Master to a recording track*.

Application:
- Switching signal sources while recording (!).
- Sending the →Metronome signal to an external device.

  * *The AUX and DIRECT Masters are located to the right of the Input Mixer on the page "EZ Routing".*

Disc at Once
→CD Burning (for audio CDs) can be carried out in "Disc at Once" (DAO) mode, which writes the entire CD in one go. (The laser doesn't stop working between tracks.) This is only useful if the mastering tracks already contain all the CD tracks, and if their sequence, blanks and →CD Markers have been set. No additional data can be written to the disc at a later stage. This approach conforms to the →Red Book Standard and could be used for medleys, etc., or for any application where control over the duration of the blanks between tracks is needed. The other approach, "→Track at Once"

(TAO), only burns one track at a time. Either method can be selected via the "CD-R WRITE" menu.

CD-R Write → Write Method

Disc Image File
The "→CD Burn" operation is based on a single, continuous file, stored on the internal harddrive. This file needs to conform to the CD format. While "→CDR"-format mastering tracks (with the "*" symbol) generated in the Mastering Room can be used in their present condition, regular V-Tracks must be converted to "Image Files" before burning them. This automatic operation requires additional hard disk capacity, which can, however, be sourced from any partition. This image file is deleted after selecting "NO" in response to the "Write another Disk?" message that appears after the burning operation.

Disk Memory Full!
This warning message means that the storage capacity of the current →Partition is exhausted. Consider using "→Project Optimize" to delete unused audio data, "→Project Erase" to dispose of no longer needed Projects, or even copying the current Project to another partition (→Project Copy).

Display
This menu on the VS-2000 allows the following synchronization parameters to be set: →Display Offset Time, →Time Display Format and →Peak Hold Sw.

Display Offset Time
During slave operation, this parameter can be used to activate an →Offset for the received →MTC time code. Such an offset may be necessary to align the master and slave's audio material, even though their (timecode) start positions differ (VS-2000: Display Parameter).

Utility → Project Parameter → Display Offset Time

Display Parameter

For reasons of simplicity, the →SMPTE time and measure counter can be displayed as absolute (actual received timecode) or relative (including the offset) values. Use the "→Time Display Format" parameter to make this selection.

Display Parameter

The "D." page only exists in the →UTILTY Menu of the VS-2000. It contains the following parameters:

- Timecode offset for slave operation:
  →Display Offset Time (p. 57)
- Absolute/relative time indication during sync slave operation:
  →Time Display Format (p. 225)
- Holding the highest level read-outs:
  →Peak Hold Sw (p. 145)

Dither

D. is able to increase the →Signal Noise Ratio by adding a special kind of white noise. Based on psycho-acoustic phenomena, it shifts extremely low-level signals to the audible range during an →A/D Conversion or →D/A Conversion. This is useful for high-quality conversions of 24-bit signals to the CD standard (16 bits). On the VS, Dither can be added to digital signals, which have to be transmitted to other digital devices with a lower →Word Length than the project's current one:

Utility → Project Parameter → Dither

Burning CDs
When burning a CD from a 24-bit project (also →MTP), the VS automatically converts them to 16 bits thus automatically dithering them.

Sending digital Signals to external Devices
When transferring →Digital Audio Signals via →S/PDIF output to a device that uses a lower bit resolution, dithering is recommended. This avoids truncating the last eight bits (whose information would otherwise be lost).

VS-24xx: *Utility → Project Parameter* → Dither*

On the VS-2000, the Dither setting can be found in the "Dig. I/O Parameter" menu.

Divide
→Phrase Divide (p. 149)

Drive
On the VS, each →Hard Disk →Partition is called a "drive" (except in the →Project List where they are called "IDE drives").

Drive Busy!
The D. warning usually means that data access is too slow due to hard disk →Fragmentation or an unsuitable hard drive.

Drive Check
If a problem occurs while loading or copying a →Project, the TOC (Table Of Content) of the current →Partition (Drive) should be checked:

*Project → select a partition →
Page → F1 DrvChk*

After checking the drive, the VS displays a report and one of the following messages:

No Err: No problems found.
xx Err: Found the following problems:

Defect Cluster:
"Clusters" are the smallest data units the VS can access on a hard disk. Cluster problems usually mean that certain passages can no longer be played back.

X-Link Err:
"Cross link errors" refer to mismatches between the audio data and the Project's TOC. This often means that the wrong Phrases are played back (sometimes even Phrases of other Projects on the current partition).

Loose Area:
Refers to clusters that cannot be allocated to any part of any Project.

Illegal Dir:
The number of faulty tables of contents.

Read Error:
Caused by faulty hard disk sectors.

Though the VS can attempt to repair damaged data, this operation often leads to data loss, due to erroneous entries being deleted from the data TOC.

Salvaging Data
If the faulty Project can't create a →CD Backup, or copy its data to another partition, *do not* try to repair it right away. First try to salvage the data of all readable tracks in one of the following ways.

- Create WAVE files on a CD using →Track Export or →Phrase Export.
- Create audio CDs based on track pairs (→CD Burn).
- Transmit the data to a DAT recorder, audio sequencer, etc., in the digital (or analog) domain.
- Locate the required Takes in the →Take Manager and use the "Preview" function to transmit them to an external device.

After salvaging all required (or still available) data, the →Formatting of the hard disk is recommended.

Drop Frame
This →Frame Rate is used for the →Synchronization to NTSC-format color videos in the US, Canada, South-America, Asia, etc. Each second is divided into 29.97D frames. Given the fraction, two frames are dropped per minute (except for the minutes that can be divided by "10", →Frame Rate).

Utility → Sync Parameter → Frame Rate

DS-30, DS-50, DS-90A
These discontinued active monitor speakers with digital inputs from Roland are used as a reference for the →Speaker Modeling →Effect, which simulates the behavior of various studio monitors but requires DS-xx speakers.

DSP
Is short for "digital signal processor", i.e. a processor capable of processing audio (or other) signals in realtime. Thanks to the separation of program and data busses as well as the use of additional arithmetic units that perform the multiplications required for audio signal processing, huge amounts of data can be processed quickly.

Dual-Function Buttons
Many buttons on the VS's operation panel have two functions. The second function can be selected by holding down [SHIFT] while pressing the desired button. This procedure provides access to the functions shown in the labelled box.

DualComp/Lim
Each →Effect block of this dual-mono dynamics processor can be used as either a →Compressor or a →Limiter. The following block is a →Noise Suppressor that acts as a noise gate. Note that the channel →Dynamics processors' quality is such that this effect will rarely be needed.

Dual Compressor/Limiter	
Effect patch:	P029
Connection:	Insert
Application:	Special

Ducking
A "Ducker" enables control of the level of one signal when using a second signal. The most common application of this is when the voice-over signal reduces the background music's level. This function can be simulated with the channel dynamic's →Key In.

DVD
Introduced in 1995, the "Digital Versatile Disc" is available in three varieties: DVD-Video, DVD-Audio, and DVD-RAM. The reading speed of DVD drives is six times higher than that of a CD drive (single speed). Writable and re-writable DVDs are called "DVD-R/RWs"

DVD Backup

or "DVD+R/RWs". The VS-2480 requires DVD-R/RWs for →DVD Backups.

DVD Backup

The VS-2480DVD can archive its Project data both to CD-Rs (→CD Backup) and DVDs. With a storage capacity of 4.7GB almost seven times that of a CD-R, DVDs are usually more efficient for Project data. The VS can only use DVD-Rs or DVD-RWs ("Minus-R") conforming to the "General" format (data or video). Operations are the same as for →CD Backup (p. 34).

Dynamic Automation
→Automix (p. 19)

Dynamic

D. refers to the difference between the highest and lowest levels of an audio signal. While the human ear has a dynamic range of 130→dB (between the hearing threshold level and the threshold of pain), the dynamic range of audio signals is often reduced for technical (transmissibility) and artistic reasons (perceived loudness). This can be achieved with a →Compressor or →Limiter that reduces or flattens signal peaks. Reducing the dynamic range has the advantage of increasing the level of the entire signal without causing distortion therefore making the signal appear louder.

Dynamics

D. refers to the control amplifiers provided on each of the VS's mixer channels*. This circuit can act as →Compressor or →Expander. Some applications call for the use of other signals to control the processor's operation (→Key In). The "Dynamics Sw" switches the processor on/off; "Dyn Typ" is used to select the compressor or expander algorithm. The dynamics processors can be configured from the "CH Edit" menu:

VS-2400: not for Input Mixer channels 9~16.
VS-2480: simultaneously use is possible.

CH EDIT → Dynamics

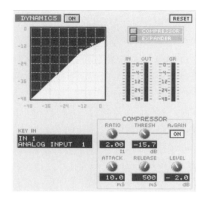

Fig.: Compressor/Expander on the VGA Screen

Tip
The "→MltChV" page displays the dynamics curves of all channels. To switch a processor on or off, click the blue dynamics field.

VS-2480
Here the expander and compressor, per channel, can be used simultaneously. This may prove invaluable for creative drum recording and mixing. The joint use of both processors is subject to the following limitation: Using both dynamics processors for one channel (e.g. on the Track Mixer) is possible. In this case, the other mixer section (here the Input Mixer) does not have channel dynamics.

E

Edit Message

Before confirming and executing an edit operation (→Editing), the time and parameter values entered can be checked in a special flip menu. This is especially useful for the mouse and associated graphic edit operations.

Utility: Global2 → Edit Message
VGA: Right-click the Playlist → Edit Msg.

Edit Point Buttons

The →Edit Points are used for most edit operations to define the passage to be processed, the target time etc. The Edit Point buttons set the IN, OUT, TO and FROM without calling up the associated menus via a mouse click in the VGA or by pressing the appropriate VS bottons. During numeric editing, the edit points can be set in a table. Though graphic →Mouse Editing generates those edit points automatically, the buttons can still be used to correct their settings.

Setting Edit Points
Locate the desired storage position, then assign it to the appropriate button:

VGA:	Right-click the →Measure Bar
VGA:	Click the ⌐ ¬ * icons, see below
VS:	press the appropriate button.
VS-2400:	Press SHIFT + button.

The Edit Control Strip is located above the measure bar and displays the timecode locations of the edit points.

Fig.: Edit Control Strip (VGA)

For operating with the VS buttons, there is an "→Edit Point Sw Type" parameter that can specify whether pressing a button overwrites the corresponding Edit Point, or simply locates it. During numeric editing, the edit points can be set in a table.

Correcting Edit Points
Existing entries can be changed by dragging the mouse, specifying a new time code location (see below), or by overwriting them (see setting Edit Points).

Locating to an Edit Point
Hold down [SHIFT] while pressing the desired button to jump to the corresponding position within the Project (not available on the VS-2000). On the VS-2400, this requires changing the "→Edit Point Sw Type" setting.

Clearing Edit Points
Though the edit points are overwritten each time they are stored to a new location, clearing them may prove useful at times.

VGA:	CLEAR + click icon.
VS:	Press CLEAR + button.
VS-2400:	Press SHIFT + CLEAR + button.

The behavior of the VS's edit button can be changed using "→Edit Point Sw Type".

Edit Point Sw Type
For the →Edit Point Buttons (IN, OUT, TO or FROM), the E. setting specifies whether pressing one of those buttons overwrites the timecode location, or whether the VS jumps to that position (VS-2000: *System):

Utility → Global1* → Edit Point Sw Type

Overwrite
By registering a new location, the previous one is overwritten. In most cases, this default setting will suffice (for the (VS-2400, see below)

Same as Locator
This setting means that the current location is only stored to any free memory space whose button doesn't light up when it is pressed. For buttons that already contain a location, first use CLEAR (→Edit Point Buttons). Unlike "Overwrite" mode, this setting means that the Project locates the position stored for the button pressed. This is similar to the behavior of the →Locator function. It protects the Edit Points from being accidentally overwritten.

VS-2400:
On the VS-2400, hold down [SHIFT] to set an edit point.

Edit Points
→Editing of audio data is generally performed based on the passages set with the IN, OUT, FROM and TO buttons (→Region Edit, →Automix Editing). These locations are temporarily available in the edit memory, can be accessed with the →Edit Point Buttons and can be set on the VS itself or on the VGA screen.

Editing

In: Region Start
Out: Region End
To: Destination location
From: Time anchor within a Phrase/Region

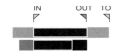

Fig.: IN and OUT Edit Points

While the E. are not really necessary for →Mouse Editing, their information is nevertheless available and can be corrected if need be (see the icons in the →Measure Bar).

The Region starts at the IN point and ends at the OUT location. In most instances, →FROM uses the same location as IN. Special applications, where a given event within (or outside!) a passage needs to be moved, copied, etc., to a specific location, may call for a different FROM setting, however. The target location for a copy, etc., is specified with the TO position. The Edit Points are graphically indicated by icons, which can be generated, moved, cleared and corrected. On an external VGA, the ▼ ▼ icons are used for IN/OUT in the →Measure Bar (in the LCD, ▪ ▪ icons).

Editing

E. refers to the electronic equivalent of tape splicing, although it includes a lot more (deleting, copying, inserting, splitting, trimming passages, etc.). The VS provides the following functions:

1. Editing Regions (between the IN and OUT points, which are define by drawing a rectangle with the mouse or entering the appropriate values):
   →Region Edit (p. 188)
2. Editing complete track areas, which were generated during recording, or separated using the split operation (track bars):
   →Phrase Edit (p. 150)
3. Editing areas between two markers. This applies to all tracks in the Playlist:
   →Region Arrange (p. 184)
4. Editing Automix events:
   →Automix (p. 19)
5. (On the VS-2480) Editing recorded Phrase Sequencer events:
   →Phrase Sequence

Editing can be performed in one of the following ways:

- With the mouse in the VGA and the LCD
  →Mouse Editing (p. 132)
- Via a table (the most precise approach)
  →Numerical Editing (p. 136)
- Using →Wheel/Button Editing (p. 243).

Effect

Effects are used both to add color and also to change the sound dramatically in the sense of creative sound design. In this way they are able to determine the sound of an entire arrangement.

The VS provides a host of effects, which are based on 37 algorithms that can be selected via the 200 presets. There are also 200 "User Patches" where original versions can be stored. The effects processors reside in independent →Effect Boards, which need to be installed in the VS (each board has two effects processors).

Model	Max. effects
VS-2480	8
VS-2400	4
VS-2000	6

By installing optional →VS8F-3 boards, third-party →Plug-ins can be added. These effect programs need to be installed on the internal hard disk. When selecting such a processor/effect, the VS loads the corresponding settings in the board's RAM memory.

Effects can be connected in one of two ways: as →Insert or →Send/Return Effect effects (depending on the need for their use). The most commonly used effects, like →Reverberation, →Delay and →Chorus, are normally used in a Send/Return configuration.

Effect

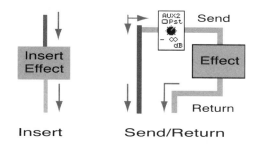

Fig.: Effect connections

Send/Return effect:
A →Send/Return Effect (or "loop effect") allows the balance between the effect ("wet") and the original channel signal ("dry") to be set. Using the →AUX Busses*, as many channels as are desired can be processed with the same effect. Depending on the number of installed →Effect Boards, several effects can be simultaneously used (e.g. AUX1~AUX4 on the VS-2400). Typical loop effect applications are reverb, chorus, and delay. The following describes the configuration of an effect as Send/Return effect and should be used as a quick start. See →Send/Return Effect (p. 204) for a detailed discussion.

> * *The VS-2000 uses the term "FX" for the internal effects (rather than "AUX").*

1. Use →CH EDIT to select the desired mixer channel.
2. Set the AUX1 (send) knob approx. to "–10dB" (by default, this bus is connected to a reverb effect).
3. Check the bus fader setting if necessary: The bus signal level can be set by an →Aux Master fader located before the processor's input. In most instances, the default setting (0dB) should be fine.
4. Check the FX Return setting if necessary: The processor's stereo output level can be set with the associated →FX Return Channel prior to being sent to the mix bus. This control sets the effect's overall level.

Insert effect:
→Insert effects are used to process just one (stereo) signal at a time. As they process the original channel signal, they are usually used for →Compressor, →Electric Guitar, →Vocoder effects, etc. linking several insert effects.

1. Use →CH EDIT to select the desired mixer channel.

2. On the VGA screen, click the Insert field of the desired effects processor. In the LCD, select the "FX INS" field and press YES. Move the cursor to the desired effect slot and activate it with the dial.

Selecting Effects
To select an →Effect Patch, first choose the effects processor (Effect1~4 on the VS-2400). via the menu on the main effects page, or directly on the VS.

VS:	SHIFT + F3 (Effect)
VS-2480:	EFFECT

The main effects page shows the number and types of installed →Effect Boards. To select an effect:

VS:	F1~F6
VGA:	Effect menu
VGA:	Double-click an effect field
VS-2000:	FX1~FX6

Use the "Algorithm View" of the current →Effect Patch to select the patch list, the edit page, →Bypass and the "Save" page:

Selecting an Effect Patch from the List:
Select the desired preset patch with the mouse, dial or ⬍ cursor. To recall the associated settings, press F5 (Select) or YES or double click on an patch name.

Effects For The Master Bus
To process the Mix Bus's signal, an →Insert effect can be assigned to the →Master Section:

Effect Algorithm View

VGA: Mixer → Master Edit → F2 (FX Ins)
VS: SHIFT + MASTER
VS-2400: MASTER EDIT
VS-2000: SHIFT + F3 (Effect)

Inserting such an effect is performed in the same way as per mixer channels.

External Effects

The VS allows work with →External Effect Processors and makes it possible to include them in the →Automix. External processors can be fed by an →AUX Bus or a →Direct Path. Use →Output Assign to send the signals of the selected bus to the desired output sockets (digital or analog). The external processor's outputs need to be connected to the desired Input Mixer channels.

Effect Algorithm View

On an effect's A. page, the desired →Effect Patch can be selected from the patch list and its associated Edit and Save pages can be called up. This is done via the VGA menu (on the VS-2000, also use the FX1~FX6 buttons).

Effect Board

The VS's →Effects are generated by an E., with each board providing two effects processors. These boards provide their own processors for generating effects and do not influence the system load. In this way, the independent DSPs ensure the VS's operation safety and on the other hand unlimited effect processing. Each VS is supplied with one pre-installed VS8F-2 board. To use plug-in effects from the third-party manufacturers like Antares, T.C. Electronic, Massenburg, Universal Audio, it is necessary to install at least one VS8F-3.

VS8F-2

One VS8F-2 board is pre-installed in each VS. It provides 250 ROM presets and 200 user patches, which are stored in the Flash-ROM. This is why →Effect Patches can be changed quickly without having to load them first (unlike plug-in effects/a VS8F-3). This board can generate standard effects (reverb, chorus, delay, etc.), →COSM effects (microphone simulations, guitar amp modeling, etc.) and even →Harmony effects.

VS8F-3

An optional VS8F-3 serves to integrate Plug-in effects. Unlike a VS8F-2, it contains no algorithms or patches: it only provides the DSP power for loaded →Plug-in algorithms loaded. Conceived as an open platform, it allows work with additional effects developed by Roland but also by other, third party manufacturers (usually famous effect algorithms previously available only as plug-ins for PC-based software). Thanks to its internal 56-bit signal processing and sampling rate of up to 96kHz, such a plug-in board (and the available software) produces high-quality results.

Note that VS8F-3 effect expansion boards need to be installed according to a set system. Their position is of the utmost importance in order to authorize the plug-ins being installed. This may involve moving already installed effects boards to other slots. To install a board, turn the VS upside down (with the AC connector facing up) and remove the cover plate (or the entire bottom panel of the VS-2400).

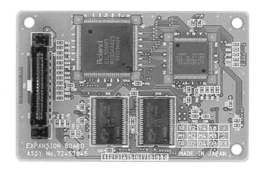

Fig.: VS8F-3

Each effect expansion board is held in place by three resin pins and clamps. To remove a board, push the clamps away from the board and carefully pull the board out of its socket. To install a board, check whether its three

Effect Patch

holes match the three pins, then push it downwards.

VS-2480:
The VS8F-2 installed in slot "A" needs to be removed and replaced by the VS8F-3. The VS-2480 allows the installation of up to four (three additional) effect expansion boards. With the exception of the first board, their order can be selected freely.

VS-2400:
The pre-installed VS8F-2 is connected to slot "A" and needs to be replaced with the VS8F-3. The VS8F-2 can then be installed in slot "B".

VS-2000:
The internal VS8F-2 is soldered onto the main board and occupies slot "A" (this cannot be changed). The VS8F-3 "Key Card" must therefore be installed in slot "B".

When adding boards (whose order is of no consequence), never leave the first slot after the "Key Card" empty.

Effect C.C. Rx Sw

→Effect parameters can be edited with →MIDI →Control Changes. Be aware that the VS doesn't transmit MIDI data when editing an effect (as is the case when changing the mixer parameters). It is imperative, therefore, to program them by entering the appropriate NRPN messages on a →MIDI Sequencer track. Such messages are only received if E. is set to "On":

Utility → MIDI Parameter →
"EFFECT C.C. Rx Sw"= On

Effect Economy

With a little planning, more →Effects can be used than the available processors can muster. During →Mixdown it may sometimes feel as if there are not enough processors for the work that needs to take place. It is therefore recommend to plan ahead and devise appropriate strategies. These could be, for example, recording track signals with →Insert effects, recording the →Effect Return signals, →Bouncing tracks with different effects to various V-Tracks etc. The →Master Effects should finally applied in a separate →Mastering operation and not during Mixing. Consider adding →External Effect Processors at the mixdown stage—or recording their output signals at earlier stages.

Effect P.C Rx Sw

To select →Effect Patches with MIDI →Program Change messages, this parameter must be set to "ON" (→MIDI Effect Patch Select).

Utility → MIDI Parameter →
"EFFECT P.C. Rx Sw"= On

Effect Patch

The 250 →Effect patches in the VS are derivated variations of the 36 →Algorithms (effect typ). The patches are named for easy identification (according to application, etc.). Before selecting an E., the respective processor must be chosen (1~4 on the VS-2400). Use the menu of the VS main effects page, or press SHIFT + F3 (VS-2480: [EFFECT] button). The main page displays the number and types of installed →Effect Boards, while it also allows →Bypassing insert effects. The illustration below shows a regularly installed VS8F-2 expansion board. Here, Effects processors 1 and 2 can be used.

Fig.: The VS-2400's 'Effect View' page

If a →Plug-in board has also been installed, the "EFFECT VIEW" at first displays "PLUG-IN" [No Plug-In] for both processors. While the VS8F-2 board's internal effects reside in a special ROM area and can be selected right away,

Effect Patch

this is not the case with plug-in boards: the desired plug-in must first be loaded into a RAM area (→Plug-in), which takes time.

Selection:
An internal effect's "ALGORITHM VIEW" page can be selected as follows:

VS:	SHIFT+F3 (Effect) → F1~F6
VGA:	Effect menu
VGA:	Double-click an effect field
VS-2000:	FX1 - FX6
VS-2480:	EFFECT → F1~F6

Press F1 to call up the effect patch list:

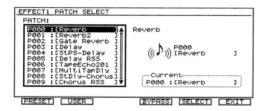

Fig.: Effect patch list

Selecting An Effect Patch From A List:
Preset patches can be selected with the mouse, the dial*, or the ⇕ cursor buttons (hold down [SHIFT] to change in steps of 10). The currently selected effect is still active at that time. To select a user patch, press F2 to call up the User Patch list (see below).

Activating It
Press F5 (Select), YES, or double-click to load the settings of the selected patch. The name of that patch now appears in the "Current" field.

Recalling Effects From Other VGA Pages
On an external VGA screen, effect patches can also be selected elsewhere:

- VGA's →CH EDIT page:
  Click the patch number of an →Insert effect (displayed in blue).
- →TR F/P and →MltChV pages:
  Click the blue FX field on the "FX Return Mixer" page.

- →FX Return Channel page:
  Click the black FX field.

Selection Via MIDI
Preset and User patches can also be recalled using "→MIDI Effect Patch Select". This is based on →Program Change and Bank Select messages.

Editing
Effects patches are edited by selecting a Preset or User patch that uses the required →Algorithm, after which the parameter settings can be changed as desired. The next step, is to save the settings to a User patch. (As an alternative, a new →Scene, which also includes the mixer settings, could be created.)

Algorithm View
This page displays a block diagram of an algorithm. Most algorithms provide several effect blocks. Some of these blocks (or "modules") can be switched on/off here using the mouse or the dial. Double-clicking a module, or selecting it and pressing F3, calls up its edit screen.

Edit Screens
The edit screens on a VGA display are similar to those shown in the LCD. So F1 and F2 are used to change to the desired page.

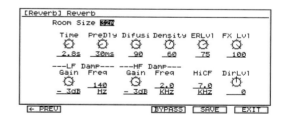

Fig.: Effect edit (Reverb, screen 2)

Parameter values can be edited by clicking them and dragging the mouse, or by selecting them and using the dial. →Bypass could be used to compare the original signal with the processed version (→A/B Comparison)

Effects, recording

Saving
Click "Save" or press F5 to call up the User patch list where the saved patch to be overwritten can be selected. The patches originally displayed in that list are copies of the Preset patches and can be overwritten. Next, press F5 or YES and confirm the question the VS displays to save the settings to the selected User patch.

Tip
Changes to effect patches are saved along with the mixer and routing settings in a →Scene. Because of this, it can be more convenient to not use User effect patches.

Entering a Name:
In the "Save" page, the new effect patch can be named. →Name Entry is an important step, because the user effect patches can be accessed from any Project.

Higher Sampling Rates
At →Sample Rates beyond 48kHz, the following internal effect algorithms are no longer available: Reverb 1, Gate Reverb, MicModeling, Voice Transformer, Vocoder 2, Speaker Modeling and Master Tool Kit. The →Plug-ins, on the other hand, also support higher sampling rates. The "→SRV Stereo Reverb" and "→Mastering Tool Kit" effects are also available as plug-ins.

Effect Plug-in
When installing the optional effect board →VS8F-3 additional Plugins (p. 159) from third party companies can be used.

Effect Return
An E. represents an effect processor's stereo output which represents the effect signal (wet). It is processed by a special →FX Return Channel prior to being fed back into the →Mix Bus.

Effects, recording
For →Effect Economy reasons, the signals output by an internal or external processor are recorded on a separate track, which leaves the effects available for other applications. This procedure is also called printing effects.

Insert Effects
Here, a track channel with the one inserted effect will be routed to a new track (→Bouncing) and the recording will be carried out. This track then contains the processed version in data form and further settings are unnecessary (useful for GuitarAmpModeling, MicModeling, Vocoder, etc.).

Send/Return Effects
Route the processor's stereo output to a stereo track and then record the (wet) signal. A mix of the effect and the original signal can also be tracked simultaneously - the second approach no longer allows the effect "depth" to be changed.

Tip:
When recording various reverb types and different wet/dry mixes (→AUX Send) onto separate →V-Tracks of a main track, in the final mix changing the effect is then a matter of switching V-Tracks. Consider the following example:

V-Tr. 1:	Long reverb
V-Tr. 2:	Medium reverb
V-Tr. 3:	Short reverb
V-Tr. 4:	Long reverb, Voc. Send= –8
V-Tr. 5:	Long reverb, Voc. Send= –4

To use those effect (V-Tracks) simultaneously, they must be copied to active tracks. (Given the number of available V-Tracks, it is strongly recommend to preserve the original "dry" versions, so as to be able to repeat the bouncing operation if the result proves unsatisfactory.)

External Effects
The output signals of external processors need to be recorded to stereo tracks (→Recording). It is recommended to record several reverb types and versions to different V-Tracks to have more options later.

Eject

Eject
Removable discs connected to the VS-2480 are ejected automatically when the system is shut down using SHUT/EJECT.

Electric Guitar
→Guitar Hi-Z, →Guitar Amp Modeling, →GuitarMulti

End
This short cut enables the last passage recorded on active V-Tracks to be located:

VGA:	Hold SHIFT + click ▶▶
VS-24xx:	SHIFT + PROJECT END
VS-2400:	SHIFT + FF

Enhancer
An E. is an effect that boosts the high frequencies of the audio signal it processes depending on the signal's level. This means that it doesn't boost hiss or noise. By contrast, an →Exciter generates new overtones that are added to the original signal. The VS provides Enhancer effects that are available as blocks in the following algorithms: →Vocal Multi, →Stereo Multi and →Mastering Tool Kit. The "Sens" (Sensitivity) parameter sets the level at which the high frequencies are boosted. "Freq" sets the frequency threshold, and "MixLvl" specifies the mix between the processed and original signals. Enhancers are usually only used for rather dark/mellow signals.

ENTER/YES
Some operations need to be confirmed with YES (and can be aborted with NO). The VS provides dedicated buttons (and fields on an external VGA screen). When a confirmation message appears, the YES button starts flashing.

EQ
Short for →Equalizer.

Equalizer
An EQ (sometimes called a "filter") serves to equalize frequencies and allows a signal's frequency response to be changed or corrected by boosting or cutting certain frequency ranges. Such changes may be applied for technical (e.g. cutting ground loop noises, etc.) or creative reasons (making the vocals brighter, etc.). Recording studios usually use →Parametric Equalizers (sometimes also →Graphic Equalizers). All equalizers use filter curves to determine which frequencies pass through unaffected and which are boosted or cut. The pass-through range (shown in white in the illustration below) refers to the frequencies that are allowed to pass. They can be left unchanged (by setting GAIN to "0dB") or to increase their level (by selecting a positive GAIN value, e.g. "+9dB" for "280Hz" in the illustration). To cut unwanted frequencies, select a negative GAIN value for the corresponding filter band. High negative values (max. –15dB) mean that the band in question is suppressed almost entirely. Smaller values allow its level to be reduced. (The human ear can hear frequencies in the 20Hz~20kHz range.)

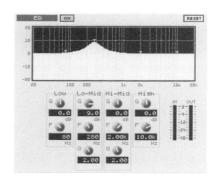

Fig.: Channel Equalizer

The entering point for the filter application is called the cutoff or center frequency (F). The frequency setting is never abrupt—neighboring frequencies above and below it are also affected resulting in a smooth transition. The Gain (G) parameter controls the level of attenuation or boosting. In peaking EQs, the bandwidth of

Equalizer

the frequency range can also be set (Q).
All VS channels of the →Input Mixer* and →Track Mixer sections are equipped with a parametric EQ comprising two shelving and two peaking filters. Each EQ can be switched on/off (ON) which means that it doesn't affect the channel signal. Via RESET all values are initialized (flat). All channel EQ settings can be automated both statically and dynamically (→Automix). By contrast, →Scenes store the EQ settings of all channels.

> * VS-2400: Only Input Mixer channels 1~8.
> VS-2480: Filter (HP, LP, BP, BEF)

VS-2400:
The EQ button in the operation panel's CH PARAMETER section calls up the "EQ" page of the last mixer channel selected. The →Attenuator at the EQ's input sets the level of the signal to be processed by the EQ. Its level is indicated by the IN meter. Be sure to also monitor the OUT level meter, because boosting or cutting certain frequencies may produce dramatic changes in volume. Its readout is identical to the channel's "Pre" indication.

Shelving EQ

The "High" and "Low" bands of the VS's EQs are so-called "shelving" filters, meaning that all frequencies above or below the cutoff frequency are affected by the G setting.

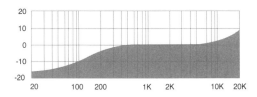

Fig.: Shelving EQ

During →Mixdown, it is recommended, to cut the frequency range below ±100Hz of all signals except the bass and kick drum by means of such filters. The level of instruments whose frequency range is increased above 6kHz seems to be moved to the foreground. Slightly reducing an instrument's "high ends" moves it into the background. Doing so broadens the "depth" of the sound image. High- and low-pass filters exhibit a behavior similar to that of a shelving EQ, except that their slopes are a lot steeper, so that the frequencies above/below them are cut rather more abruptly (only available on the VS-2480).

Peaking Filter

The two middle bands of the VS's channel equalizers are so-called peaking (bell) filters. The desired frequency band (center frequency "F") is limited on both sides (see "Q"). In contrast, shelving EQs affect all frequencies above or below the center frequency.

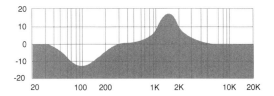

Fig.: Peaking Filters at 100Hz and 1.5KHz

Q

A peaking filter's Q ("Quality") specifies how many frequencies in the vicinity of the selected center frequencies should be boosted or cut along with it. A setting of "1.4" means the range from about one octave below, to one octave above the center frequency can be boosted or cut (a setting of F= "1kHz" would therefore affect the 500Hz~2kHz range). The lower the value, the more octaves are affected (large bandwidth). Extremely high Q values, on the other hand, mean that only the selected frequency (e.g. only one note) is boosted or cut.

Tip

The →Effect Patches P186~P214 provide useful starting points for frequently-used parametric and graphic EQ settings. Selecting the same settings for a channel EQ can prove a major time saver.

Erase

High- and Low-pass filters of the VS-2480
Only the VS-2480 provides an additional filter that can be used as LPF, HPF, BPF or notch filter (BEF). The cutoff/center frequency can be set via the →Channel Strip and encoders. A high-pass filter (HPF) only attenuates frequencies below its cutoff frequency (at 12dB/octave), while the range above passes through unaffected. This filter can be used to avoid "muddying up" the sound image (leaving more room for the bass guitar and kick drum). Low-pass filters (LPF) attenuate the level of frequencies above its cutoff setting. Hardly ever used for recording or mixing applications, they may come in handy for extremely noisy signals. The →Massenburg EQ provides an LPF and a HPF with a selectable slope of up to 24dB/octave.

Cutoff Frequency And Resonance
The VS-2480's LPF filter can be used for manual (synthesizer-like) filter sweeps. Its Q parameter can be used for a resonance-like emphasis of the cutoff frequency (F), which can be changed in real-time to create filter sweeps. Such dynamic changes can be recorded using the →Automix function.

Band-Pass On The VS-2480
A B. (BPF) is, in fact, the combination of a high-pass and a low-pass filter: it cuts high and low frequencies, while the range around the center frequency passes through unaffected. This filter can be used to separate instruments and to create "frequency windows" during a →Mix. The →McDSP ChromeToneAmp plug-in's "Dist BP" mode simulates the band-pass operation of a Wah pedal.

Notch (BEF) Filter On The VS-2480
The band eliminate filter (BEF, also called "band reject" or "notch filter") suppresses a small frequency band, while all other frequencies pass through unaffected. This can be used to suppress ground loop noise, other narrow-band noises, or (at the highest Q setting) for creative applications, like cutting the 300Hz range of a kick drum.

Erase
The E. function allows the removal of unwanted data from the specified region without changing the locations of subsequent audio material. This operation corresponds to erasing a given part on an analog tape by recording silence. The VS provides E. functions in →Region Erase, and for →Automixes. Using →Cut rather than E. lets the subsequent passages be moved to the left.

Tip
The →Exchange function can be used more easily for erasing an entire track.

Error Level
In slave mode, when used for →Synchronization, E. determines how lenient the VS should be in the events of MIDI timecode (→MTC) errors. Selecting "10" means that timecode gaps of up to three seconds are accepted. Be aware, however, that higher tolerance settings also produce stronger fluctuations, which may cause problems at some stage.

Utility → Sync Parameter → Error Level

Error Message
Error messages are displayed whenever the requested operation cannot be executed. The message usually details what is impossible. Here are a few examples:

Found Illegal Track Pair!:
Regular tracks or Phrases are trying to copy to a →CDR created in the →Mastering Room, etc., which is impossible.

Can't Record CD
The →CD Digital Rec parameter hasn't been switched on.

Lack of Events
→Events

Not 44,1kHz Project!
The →CD Burn operation is only available for Projects that use the 44.1kHz →Sample Rate.

Drive Busy!
The hard disk's →Fragmentation is too significant to continue.
Boot Condition: →Skip Project Load (p. 207)

Event
Aside from →Audio Data, the VS uses Events to carry out operations. Moving a →Phrase requires two events, →Region Move uses four, and →Automix operations require even more events. Each Project can hold up to 30,000 events. When that number is reached, the VS displays the →Error Message "Lack of Events!". This means that no additional operations can be performed, irrespective of the hard disk's remaining storage capacity. To bypass the problem, save the Project, which discards the Redo data (→Undo), use →Project Optimize, or delete unnecessary Automix data. The number of remaining events can be displayed via →Remain Display Type:

> Utility → Global Parameter2 →
> Remain Display Type = Events

Exchange
→Track Exchange (p. 227)

Exciter
An E. generates special distortions that add upper harmonics to the signal being processed. Expert use of this effect adds more sparkle to a signal, which a conventional EQ usually cannot do. An E. is especially effective for protecting the high range of the original signal before processing it. An E. is chiefly used for the restoration of old recordings, to add brilliance to stringed instruments, and to enhance vocal recordings (→Effect). The VS doesn't contain an exciter effect.

EXIT/NO
Most operations can be aborted with E. (and confirmed with ENTER). Either press the corresponding button on the VS's operation panel, or click the designated fields on the VGA screen.

Tip
Whenever the YES button starts blinking, the operation can be aborted by pressing EXIT.

Expander
An E. lowers the level of signals below the →Threshold value, leaving levels above that value unaffected. This exaggerates the →Dynamic of the signal being processed. Contrast this with a →Compressor that actually compresses the dynamic range. An E. is usually used to reduce background noises (at high Ratio settings, the E. works like a →Gate).

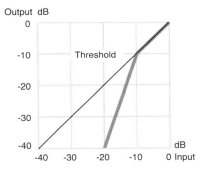

Fig.: Expander

Threshold
Only levels below the Threshold value are attenuated. Levels above that value are outputted at a nominal level (input level= output level). In the "Expander" illustration, the Threshold has been set to −10dB.

Ratio
The →Ratio parameter specifies how strongly the input signal is attenuated. Low Ratio value produces slight level attenuations, while high values result in significant attenuation. Signal levels above the Threshold are "processed" with a Ratio of "1 : 1", meaning that they don't change.

Attack and Release
The Attack parameter delays the attenuation once the signal level drops below the Threshold value (common settings: 0.5~1ms). Re-

Export Wave

lease, on the other, allows the level reduction to be maintained for a short while after the signal level has exceeded the Threshold. This may prove invaluable when the E. has a tendency to "shake".

Gate

A special variety of the E. function is called "Gate".

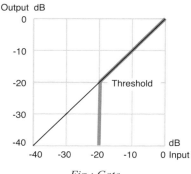

Fig.: Gate

The Ratio= ∞ setting completely suppresses levels below the Threshold. In the illustration, a –21dB signal is not outputted whilst signals with a level of –19dB and beyond are.

Channel Expander

All of the VS's mixer channels are equipped with a →Dynamics section[1] (→CH EDIT). However, only the VS-2480[2] allows the Expander and Compressor to be used simultaneously. Clicking EXP only selects the desired algorithm. To activate the expander, select Dynamic Sw= ON.

[1] *VS-2400 Input Mixer: Channels 1~8*
[2] *VS-2480: Only in one Mixer Section.*

Key In

The signal of another channel that is used to control an expander or compressor is called "Side-Chain" or "Key In". This function is mainly used for signals with a minimal signal-to-noise ratio, or for creative applications (Ducker, etc.) →Key In (p. 100).

Export Wave

The VS provides the following functions for using the material recorded on the VS in other devices or a computer: →Track Export and →Phrase Export. These functions generate Wave files that can be burned onto a CD-R.

EXT

This setting means that the VS is used as →Synchronization slave and therefore follows an external device's timecode (→MTC). This may be necessary to ensure that both the master and the slave run in perfect sync. The VS only follows an external timecode if its internal timecode generator is deactivated (→INT):

UTILITY → SYNC Prm → Sync Mode = EXT

The VS-2480 provides a special EXT button.

Ext Level Meter (MB-24)

The optional MB-24 meter bridge for the VS-24xx displays the levels of the selected signals. It contains 24 LED meters. This bridge is useful for →Live Recordings and →Live Mixing. The MB-24 has been discontinued. The meters can be configured as follows:

Utility → System1→ Display Section

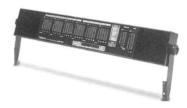

Fig.: Optional meter bridge for the VS-24xx

It is recommend to switch the meter bridge off when it's not in use or when working with external →MIDI devices.

EXT SYNC

The "EXT SYNC" indicator lights up when the VS is synchronized to an external device (→EXT mode).

External Effect Processors

Ext Timecode

Recording digital signals, sent by an external device which acts simultaneously as the Timecode →Synchronization's Master, additionally requires the Slave's →Word Clock to be synchronized. This varies the Slave's →Sample Rate depending on the Master's →Timecode fluctuations:

> Utility → Project Parameter:
> Master Clock = Ext Time Code

VS-2000: "Digital I/O" parameters of the Utility menu ("Display Parameter" section).

External Effect Processors

Thanks to the VS's flexible →Output Assignments, working with external →Effects is easy. Simply assign the desired Aux bus or DIR Path to the appropriate output. One signal path (to or from the processor) can be established in the digital domain.

Analog Connections

See the illustration below for the required connections. They are the same for external →Insert and →Send/Return Effects. Connect the VS's assigned →AUX OUT socket to the effects processor's input. Next, connect the processor's stereo outputs (available on most units) to two VS inputs.

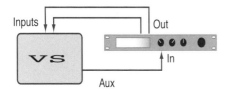

Fig.: Connecting an external effects processor

Digital Connection

Most processors don't allow a simultaneous digital input and output from an Effects device (→Word Clock).

Using the processor's digital input:
Connect the desired digital output of the VS to the effects processor's digital input. On the →EZ Routing page, route the bus being used to the selected digital output (→Output Assign).

Using the processor's digital output:
The advantage of this approach is in the effective use of the limited analog inputs. Connect the processor's digital output to one of the VS's digital inputs. Set the VS's →Master Clock to "Digital" (→Recording of digital Signals).

Routing

The →AUX Bus* called upon to send signals to the external processor must be routed to an AUX output (→Output Assign). It is recommended not to assign an AUX bus that is already feeding an internal effect. (Though possible, this would mean that channels feeding that bus are processed by two effects.) In most instances, AUX busses 7/8 will probably be used.

> *For stereo operation, two busses need to link in the →Master Section.

In the illustration below, the signals shown left are transmitted to AUX busses 7/8, the node establishes the connection to the AUX A sockets.

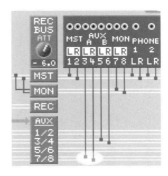

Fig.: Routing busses 7/8 to AUX A

Send/Return Effect

The settings discussed above mean that AUX 7 sends its signals via AUX socket 3 (AUX A) to the processor's input. The processor's outputs (→Effect Return signal) are connected to two VS inputs whose input sensitivity needs to be set (like for →Recording signals). This means that the processor's output signal is fed to the

EZ Routing

→Input Mixer where it can be set and processed (if necessary). See →External Effect Processor via ADA-7000 (p. 172) for how to connect an external effects processor to the R-Bus port.

Insert Effect

To use an external processor as →Insert effect, establish the same connections as per a Send/Return operation. As only one channel is processed here, it is recommended to work with a →Direct Path.

1. Activate the DIR 1 (or any other odd-numbered DIR) button of the desired channel.
2. Deactivate the →MIX Switch.
3. On the EZ Routing's output page, route DIR 1 to AUX A.

This connection means that the selected channel's send signal bypasses the mix bus and goes directly to the DIR 1 bus, and reaches the external effect through the selected AUX socket. (compressor, distortion effect, etc.). Its outputs are sent to the →Input Mixer.

Tip:

- The Input-Mixer's faders are generally used for leveling the processor's effect return signal. It is more convenient to establish leveling via the Track Mixer fader that is no longer needed (the one belonging to the channel being processed). Source the channel's DIR signal "Pre" fader, then create a →Fader Group that includes the described input fader.
- A→Master Insert with External Effect (p. 116) can also be used.

EZ Routing

The EZ-R. allows any signal to be routed to any destination inside the VS as well as to its physical inputs and outputs. All connections are established using virtual patch cables (that can be "connected" with the mouse, the cursor buttons, the dial, and/or the Status/Select buttons). The VGA displays all sections on a single page with a photo-like display of the input and output sockets. The LCD, on the other hand, is spread over several screens.

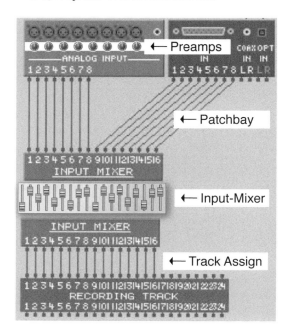

Fig.: Patchbay and Track Assign on the VGA Screen

1. Patchbay

A section (or device) that allows the assignment of inputs sockets to the →Input Mixer channels is called a "patchbay". Because of the switchable →Phantom Power function, the VS-24xx's LCD uses a separate page. All other assignments can be set on the "VIEW" page, however. On the VS-2000, only this page serves for viewing the information—Routings can't be changed here.

Stereo Channels

- All patchbay cables (between the inputs and the Input Mixer) are configured as stereo pairs. This cannot be changed.
- Linked mixer channels display only one connection node.

In the VGA display, the Input Mixer is split into two rows. The upper row represents the inputs, while the lower row contains "outputs" that can be connected to the desired recording

EZ Routing

tracks. The channel strips with the faders are located between them. Connections can be established and broken in the following ways:

Mouse:

- Click a node in the upper Input Mixer row.
- Hold down the mouse button while dragging the mouse up to the desired input. Release the button when the cursor's shape changes to 🖑.
- *Canceling*: drag towards the right and release.

Cursor Buttons and Dial:

- Press F2 (P.Bay) to select the patchbay.
- Select a mixer channel with ◀▶.
- Use the dial to choose the desired input. *Canceling*: turn the dial towards the left.

2. Track Assign

The next step is to connect the selected Input Mixer channel Out to the desired recording track. This operation establishes the signal flow from the Input sockets to the Recording Tracks.

Mouse:

Click the appropriate Input Mixer output and drag a patch cable to the desired track number.

Dial:

Press F1 (TrAsgn) to move the cursor to the "Track Assign" field where the required Input Mixer → Recording Track connection can be established. Use →Cursor Buttons ◀▶ to select the desired Input Mixer output, and the dial to assign it to the track.

Quick Routing Via Status/Select Buttons

The Status/Select button allows a direct routing of tracks to the input channels. This method is especially useful for the VS-2000. The starting point here is the desired recording track:

VS-24xx:

- First, activate the desired track section using the →Fader Buttons (e.g. TR 1~12)

- Press and hold the STATUS button, corresponding to the desired Recording Track for more than a second to call up the "Quick Routing"* page. The buttons start flashing to signal that the assignment can proceed.
- Press the Fader button of the desired Input Mixer section to activate it (usually the first section, i.e. IN 1~12 on the VS-2400).
- In this mode, the STATUS buttons flash green to signal that a track can be selected, while the SELECT buttons flash orange, when an input channel can be selected.
  Example: If STATUS button 3 lights green and SELECT button 5 lights orange, input channel 5 is connected to recording track 3.

VS-2000:

- Activate STATUS mode (the buttons light red) and hold down one button* to jump to the "Quick Routing" screen.
- In this mode, the STATUS buttons flash green to signal that it is possible to select a track. The SELECT buttons flash orange, indicating that an input channel can be selected (see above).

* *The time it takes to call up the screens can be set with the "→Switching Time" parameter.*

Canceling Routings:

Patch connections can be cleared by pressing the corresponding SELECT button again. To clear all connections, press F4 (AllClr). This updates the "EZ Routing" environment's graphic display.

Quitting

Press the [EXIT] button to leave the "Quick Routing" page.

3. Bouncing

The term "→Bouncing" refers to the internal recording of one or more tracks (or →FX Return Channels or →AUX Busses) to other tracks. It works in much the same way as the procedure described above. To clear a connection, drag the unwanted cable to the far right.

F Buttons

Tools

Both the patchbay and the track assignments can be cleared or initialized (reset to the factory defaults). First, press PAGE to select page 2/2. The →Function Buttons F1~F4 select the desired function (VS-2000: F3 and F4).

4. Output Assign

F3 (Out) moves the cursor to the "→Output Assign" screen:

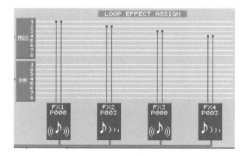

Fig.: Assigning AUX busses to effects processors

F4 (Effect) moves the cursor to the "Loop Effect Assign" screen. Next, use the mouse, the cursor buttons, and the dial to change the desired routings.

6. Templates, Loading and Saving

F5 (Load) allows to load →Template preset or own User Routings for →Recording, →Mixing, →Bouncing and →Mastering. Unlike →Scenes, which also include the routing settings, the EZ Routing settings are independent of a given Project and can be used in several Projects. Routing settings can be exported as →SMF files. This allows them to be recorded with an external →MIDI Sequencer, or to be shared with other VS owners.

There are 20 memory places where individual routings can be saved. Press F6 (Save), select an empty memory location, enter a name (→Name Entry), and finally press YES to save.

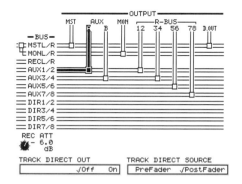

Fig.: Assigning outputs on the VS-2400 (LCD)

The →AUX Bus and →Direct Path signals (horizontal rows) can be patched to the desired outputs (columns). In the illustration above, AUX bus 1/2 is routed to the AUX A OUT (3/4) sockets. Activating the →Direct Outss is a matter of setting "Track Direct Out" to "ON".

5. Effect Busses

By default, the →AUX Busses are assigned to the →Send/Return Effects of the same number.

Aux-Bus 1 → FX 1
Aux-Bus 2 → FX 2 etc.

In most cases, these assignments will remain unchanged. The VS-2000, on the other hand, doesn't even allow them to be changed. After installing a →VS8F-3 effect expansion board, however, the assignments can be changed. To maintain the former FX1 and FX2 Send settings AUX1 and 2 can be routed to effects 3 and 4.

F

F Buttons
→Function Buttons (p. 81)

Fade
→Phrase Parameter (p. 154)

Fade Curve
The level envelope for the →Crossfade with →Phrase Parameter can be either exponential or linear. The →Automix environment's →Gra-

dation parameter also provides several curves to choose from.

Fade Length
→Crossfade (p. 50)

Fader
A F. (sometimes called "slider") is usually used for setting the level of channel and bus signals. They are able to either decrease (negative dB value) or increase (positive dB value) the Signal level. Choosing the "0dB" position means that the level is left unchanged (→dB). Digital mixers usually provide more mixer channels than faders. Therefore it is necessary to assign the Faders to the respective →Mixer sections. (example: Track Mixer channels 1~12 or 13~24 on the VS-2400). This is what the VS's →Fader Buttons are for.

VS-2000:
The VS-2000's faders (which are not motorized) only allow the track levels of the →Track Mixer to be set. They can't be used for any other purpose. The levels of the →Input Mixer channels need to be set via the display.

VS-2400:
Unlike the VS-2480, the VS-2400 doesn't provide additional PAN/AUX knobs. Both the Pan and AUX SEND values are indeed set using the faders.

Fader Functions
In addition to the level, the motorized faders (not available on the VS-2000) can be used to set the following parameters:
- Setting the L/R position of channel signals: →Panorama (p. 141) (VS-2400)
- Setting the mixer channels' Aux Send levels: →Faders for AUX Send Levels (p. 78)
- Setting the effect level: →FX Return Channel (p. 82)
- Transmitting MIDI control changes: (→V.Fader (p. 235))
- Editing an assigned mixer parameter: →Knob/Fader Assign (p. 101)

Fader Display
The LCD and an external VGA screen can display several faders simultaneously. Special menus provide a clear overview of the fader positions and even allow them to be set using the mouse or dial.
- VGA display of mixer parameters and graphic indications for the EQ, dynamics and surround settings: →MltChV (p. 130) (not on the VS-2000)
- Track Mixer: →TR F/P (p. 226)
- Input Mixer: →IN F/P (p. 96)

Fader Link
"Fader Link" allows to set the level of two adjacent channels by moving just one fader. This is convenient for stereo signals. By contrast, the →Channel Link function applies to all mixer parameters of the two channels.

CH EDIT → FDRLINK

Even after establishing a stereo pair, the level of the two channels can still be individually changed if need be. To do so, click "Submenu" at the bottom of the VGA screen, or move the cursor to the level value in the LCD and press [YES]. In the flip menu then, change the setting of either channel. For →Automix, linked channels generate special "Offset" data in contrast to the "Level" data of mono channels.

Pre- and Post-Fader
Audio signals to be sent to an →AUX Bus, →Direct Path or →Level Meter can be sourced either before or after the fader (→Pre/Post).

Fader Assign
On the VS-2400, the faders can be used for setting a multitude of parameters (→Knob/Fader Assign (p. 101)).

Fader Buttons
The VS-24xx models, F. assign the desired

Faders for AUX Send Levels

mixer sections to the motorized faders. This is necessary, because digital mixers usually provide more mixer channels than there are faders. When the desired level is assigned to the fader, the VGA screen simultaneously displays the associated page: →MltChV (p. 130) (not on the VS-2000), →TR F/P (p. 226) or →IN F/P (p. 96).

Fader button (VS-2400)	
Input Mixer:	IN1~12 and IN 13~16
Track Mixer:	TR1~12 and TR 13~24
AUX Sends/Returns:	AUX1~8 FX1~4
Pan:	PAN
MIDI faders:	V.Fader
Fader Groups:	GROUP1~12

With exception of the VS-2000 a chosen mix parameter can also be controlled via the faders (→Knob/Fader Assign).

Faders for AUX Send Levels
The VS-24xx models can set the AUX and effects send levels using the faders (→Knob/Fader Assign (p. 101)).

Fader Group
→Group (p. 87)

Fader Link
→Link (p. 104)
→Fader (p. 77)

Fader Match (VS-2000 only)
After loading a different Project (→Project Select) or →Scene with the VS-2000, the fader positions usually no longer correspond to the level settings in effect. F. specifies how moving the faders at that stage should affect the level. In the illustration, the fader has been lowered to its minimum setting. After selecting another Scene, the level jumps to 0dB. The fader is then slid slowly all the way up and then down again. The channel's level changes according to the pattern shown in the illustration. Use the System menu's "Fader Match" parameter to select another fader mode:

Utility → System → Fader Match

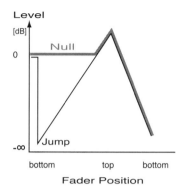

Fig.: Fader Null and Jump on the VS-2000

Jump
Any change to the fader's position immediately affects the channel's level. This may cause unpleasant jumps when the value and the fader position are far apart. Nevertheless, this setting is used by default, as any change is immediately reflected by the level of the channel being controlled.

Null
The level starts changing once the fader reaches a position that corresponds to the level value currently in effect. This means that the fader appears to do nothing for a while. Use "Null" in an →Automix and for gradual transitions from the stored level value to the desired level.

Filter
Filters are parts of →Equalizers. The VS's channel EQs provide two shelving and two peaking filters.

FILTER on the VS-2480
The VS-2480 provides an additional filter that can be used as LPF, HPF, BPF or notch filter (BEF). See →Equalizer.

Finalize
During a →CD Burn operation, F. refers to "closing" the CD to make it playable on a conventional CD player (→CD Audio). This oper-

ation copies the temporary TOC (PMA) to the disc's official TOC ("Table of Contents").

CD-R WRITE → Finalize

This operation can also be performed after burning a CD (→Track at Once).

Flanger

A F. generates a metallic, resonant sound. At extreme settings, this effect is similar to the sound of a jet plane taking off. Typically used for guitars and bass guitars, it also adds a special effect to drums (cymbals and toms), or the stereo bus.

Active Principle:
These →Effect types are based on phase cancellations obtained through splitting the signal and delaying the copy by up to 10ms (→Comb Filter Effect). In addition, a low-frequency oscillator periodically changes the offset between the two signals, creating swells of amplifying and canceling frequency components. This is precisely what causes the jet-like impression.

The following →Effect Patch on the VS generate flanger effects: →St Flanger (p. 210) and →AnalogFlnger (p. 16). A similar, though much smoother, effect can be obtained with a →Phaser.

Flash ROM

A "ROM" (Read-Only Memory) is a special kind of memory-chip whose contents cannot be changed by the user. In return, it doesn't require a power battery to buffer its content (unlike a RAM, or "Random Access Memory" chip). The F. area's contents, on the other hand, can be changed by the user. The VS uses its F. to store the →System Software, →EZ Routing templates and the User →Effect Patch of an →Effect Board.

Foot Switch

Some operations can be performed with a foot switch, which may prove invaluable for guitar players used to recording their music by themselves. Another application for a footswitch would be to set up the VS as far away as possible to avoid the unpleasant hard disk and ventilator noises. The foot switch then acts as a small-scale remote control.

Play/Stop:
Starting and stopping playback & recording.

Record:
Defeating and activating recording during playback →Punch In/Out (p. 170).

Tap Marker:
Generating →Marker (p. 109) at the current position.

Next/Previous:
Jumping to the previous or next marker, to the beginning, or the end – depending on the selection for →Previous/Next Switch (p. 165).

GPI:
Starting and stopping playback or recording using →GPI (p. 86) signals.

| VS-24xx: | Utility → Global1 → Foot Sw Assign |
| VS-2000: | Utility → System1 → Foot Sw Assign |

Formants

F. characterize the unique sound of acoustic instruments as well as the human voice. This spectral color is determined by its resonance, generated by a set of emphasized frequency bands (the body of an acoustic instrument, the mouth and throat of the human voice, etc.). F. have no influence on the perceived pitch, and are not transposed. It follows that natural-sounding transpositions require special algorithms. The VS provides the →Effect algorithm →Voice Transformer for independently transposing the pitch/formant of mono signals. In the VS-2000, the →Harmony function creates chords, which are based on the Voice Transformer, in realtime.

Format

Format
To guard against severe →Fragmentation, it is recommended to format the VS's hard disk at least once a year.

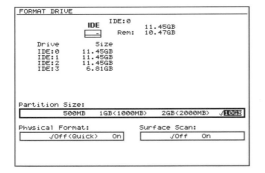

Fig.: "Format Drive" in the Project Menu

Formatting the hard disk deletes all audio and setting data, even the →Plug-in board settings. It is therefore recommended to back up Projects and uninstall plug-ins beforehand.

1. Select the "Format Drive" menu:

   Project → Project List → select IDE:0 → PAGE 4/4 → F2 "FmtDrv"

Set the parameters as follows:

Partition Size:
"10 GB" provides the longest recording time

Physical Format:
"On" deletes all data and creates a logical track and sector structure ("low-level format"). At the same time, a new directory is created, while all hard disk areas are set to "digital zero" ("nulled").

Surface Scan:
"On" means that writing and reading tests are performed after the format operation. Damaged sectors are excluded from the available storage capacity.

2. Operation
Press F5 (OK) and confirm the questions that are displayed to start formatting the hard disk. This operation may take several hours. When it is finished, consider performing a →System Update.

Found Illegal Phrase Pair!
This error message is displayed when no track has been selected for the selected →Phrase Edit operation (→Track Select).

Found Illegal Track Pair!
This error message appears when simultaneous applications of tracks in Project and →CDR format are attempted (→Record Mode).

Fragmentation
After extensive use and countless recordings, new audio data can no longer be written consecutively to the hard disk. Instead they are stored wherever possible on the HD's partition. This forces the hard disk's heads to carry out further positioning operations, which slows down access time and causes the audio data to be read more slowly. This in turn leads to a decelerated navigation within a Project, to the "Drive Busy" message, in extreme cases to clicks and pops, and even problems during →CD Backup operations.

Remedy

- Decrease the Project's size using →Project Optimize (p. 168)
- Copy the project to a less fragmented partition (→Project Copy (p. 166)).
- The best approach would be to format the entire hard disk so as to erase all data of all partitions. Before even considering this option, all valuable data should be backed up and →Plug-ins uninstalled.

Note
The HD's partitions cannot be formatted separately.

Frame Rate
The number of frames per second of →SMPTE and →MTC timecode signals is called the "frame rate". This term was created in the film industry, which uses 24fps (frames per second)

for movies. For video applications, Europeans use 25fps, while users in the US, Canada and Asia use 30fps for black and 29.97D and 29.97N fps for color video (→Drop Frame). The →Synchronization of two ore more devices requires an *identical* F. Providing the current project won´t be used for a film (24fps) score, or →Video Recorder Synchronization (25fps), consider working with 30fps because of its higher resolution. This can be set in the VSs synchronization menu:

Utility → Sync Parameter → Frame Rate

A further subdivision of the frames into →Subframe allows a →DAW more accurate positioning. With the exception of the VS-2480, all VS models are fitted with a →Sync Auto mode where the →Frame Rate of incoming MTC signals is detected automatically and therefore doesn't need to be set.

Frequency
The F. represents the most important unit of an oscillation and refers to the number of undulations per second. The measurement unit is called "Hz" or "kHz" (→Fig.: Note frequency table (p. 82)).

Frequency Response
F. refers to the band between the highest and lowest frequencies a device can output and recognize. On digital audio systems, the frequency range is determined by the →Sample Rate. The table below lists the VS's frequency ranges for the various sample rates.
The human ear can sense frequencies between 20Hz and 20kHz.

SR	Frequency range		
96.0kHz	20Hz	-	40kHz
88.2kHz	20Hz	-	40kHz
48.0kHz	20Hz	-	22kHz
44.1kHz	20Hz	-	20kHz
32.0kHz	20Hz	-	14kHz

FROM
While the IN and OUT →Edit Points specify the size of the area to be edited, TO and FROM refer to specific positions. In most instances, FROM uses the IN value, and therefore the beginning of the selected region. FROM can be used to copy passages that start with an upbeat. Activating the →Grid and shifting FROM to even beats allows arrangements to be quickly changed, or rhythmical errors to be corrected.

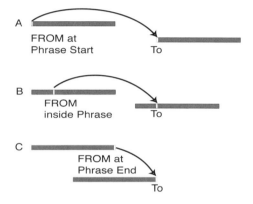

Fig.: Applications for FROM

In the illustration, example "A" represents the default situation (IN= FROM). Example "B" shows the described upbeat. "C" shows the correct position for reverse playback effects.

Note:
FROM can also be set to a position outside the IN-OUT Region.

Phrases
Moving a →Phrase, which already has a defined length, therefore only requires that the FROM and TO positions be specified. For →Mouse Editing, be aware that the cursor position, if clicked, sets the FROM value.

Function Buttons
Unlike the VS's other buttons, the F buttons don't have set functions. Their definitions change according to the selected menu, and are therefore displayed in the bottom row. Exam-

Fundamentals

ple: In the Utility menu, F4 first serves to select the "MIDI" page. Once it is displayed, F4 allows jumping to the "Bulk Dump" page.

If a menu contains more than six pages, they can be assigned to the F buttons using [PAGE]. Alternatively, instead of pressing the F buttons, use the mouse to click the corresponding fields on the LCD or a VGA screen.

Tip

By unexpected F button functions, [PAGE] usually selects the desired pages.

Fundamentals

According to the Fourier analysis, each sound is in fact a series of sine waves (harmonics) whose fundamental specifies the frequency. That is why knowing the fundamental frequency can help make efficient use of a filter. In practice, some notes are louder than others, especially on acoustic instruments. The note frequency table below allows the frequency of various notes to be determined. These levels can then be boosted using a narrow filter.

	ctr.	gr.	sm.	1	2	3	4
C	32,7	65.4	131	262	523	1046	2093
C#	35	69	139	277	554	1109	2217
D	36.7	73.4	147	294	587	1175	2349
D#	39	78	156	311	662	1245	2489
E	41.2	82,4	165	323	659	1318	2637
F	43.6	87,3	175	349	698	1396	2794
F#	46	93	185	370	740	1480	2960
G	49	98	196	392	784	1568	3126
G#	52	104	208	415	831	1661	3322
A	55	110	220	440	880	1760	3520
A#	58	117	233	466	932	1865	3729
B	61.7	124	247	494	988	1975	3951

Fig.: Note frequency table

Songs in E major, for instance, will contain the lowest bass frequency 41Hz. Bear that in mind when starting →Mixing and →Masteringing of such a song. The same is true of the octaves above and below the basic frequency.

FX

Short for →Effect (p. 62).

FX Bus

On the VS-2000, an F. refers to one of the six →Busses which feed the internal effects processors (→Effect). The mixer channel signals (→Channel Strip) can be sent to the →Bus using the →Send knob, and on to the corresponding processor's "FX Send Master" knob.

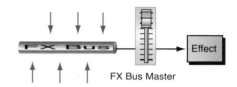

FX bus on the VS-2000

FX Send Master:

The "FX Send Master" knobs allow the bus level to be set just before the effects processor's input. By default, these parameters are set to "0dB", which normally doesn't need to be changed.

VS-2000: SHIFT + CH EDIT/MASTER

The VIEW (F1) page contains the master controls of the FX and AUX busses as well as the →Direct Paths.

> *An FX bus is identical to an →AUX Bus on the VS-24xx when the latter is routed to an effect (FX).*

FX Return Channel

The F. is a special →Channel Strip for setting the effect output signal (→Send/Return Effect). The associated fader is used to set the →Effect Return level in the →Mix Bus, and hence the

FX Return Channel

effect level in the mix. The VS provides separate return channels for each effects processor (VS-2000: 6; VS-2400: 4; VS-2480: 8). Level and Pan of F. can be automated via →Automix. The →Insert effects don't use the F.

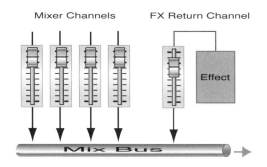

Fig.: FX Return channel

FX Return channels can be selected as follows on the various VS models:

VGA:	Mixer → CH View (FX Rtn)
VS-2480:	TR17~24/FX RTN → CH EDIT x
VS-2400:	AUX1~8/FX1~43 → CH EDIT x
VS-2000:	Hold down FX button during 1s or press F2

Except for the VS-2000, all models allow the Return level to be set using the faders.
While an F.'s structure is similar to that of a mixer channel, its function is more limited. It provides →Solo, →Mute, →Automix, →AUX Sends, →Direct Path, →Surround (not on the VS-2000) and the possibility to select →Effect Patch, but no →Dynamics or →EQ parameters.

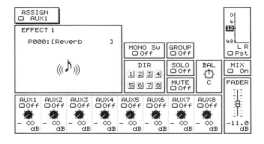

Fig.: FX Return channel (LCD)

Selecting Effects Patches:
Clicking the black effect field on the VGA calls up the corresponding →Effect Algorithm View where →Effect Patch can be selected.

Loop FX Assign:
Even though the Return channels are not related to the effects processor's inputs, a bus can still be assigned in this menu (not available on the VS-2000). This function is identical to the corresponding →EZ Routing operations. Use the mouse or dial to route a bus to the effects processor's input. The LCD's "Assign" field could also be used.

Tip
- Without changing the Send settings of the mixer channels, various alternative effects from different processors can be selected, and thus several sound options can be compared.
- After the installation of a VS8F-3 (→Effect Board), the effect assignments for the internal effects no longer correspond. For such Projects, "Loop FX Assign" can easily change the VS8F-2 boards' bus assignment (AUX bus 1 to FX3, AUX 2 to FX).

Mono
The effects processors' Return signal is in stereo, which is especially effective for reverb and chorus effects. Use the Mono setting to create special effects for Mix and the →Mono Compatibility.

Solo Modes and FX Return Channels
Activating Solo mode for a mixer channel also mutes the FX Return channels. To check a signal with effects, the corresponding FX channel(s) must be soloed as well. (On the VS-2000, →FX Rtn SOLO ENABLE can be used instead.)

Layered Effects Assignments
Effects can be further enhanced by processing them with other effects. For example, a delay effect can be processed by a reverb algorithm or a reverb can become more spacious using a

FX RTN

chorus. This routing can also be used to add a reverb effect to the →Headphone Mix. In the following example, a delay (effect 2) is sent to the reverb processor (effect 1):

1. Set the desired level for the AUX 2 Send in the desired mixer channels
2. Select FX Return channel 2 (Delay)
3. Activate the AUX 1 Switch
4. Set the level of AUX 1 (Reverb)
5. If necessary, deactivate the MIX switch to ensure that the delay signal is only output by the reverb processor.

This example can be varied for other applications. Depending on the number of →Effect Boards in use, this should allow long effect chains to be created.

Recording Effect Return Signals

To maintain an →Effect Economy, the →Effect Return signals can also be recorded. This operation is described as →Bouncing (p. 30) or "Printing Effects". This explains why the →EZ Routing environment also provides FX RTN fields.

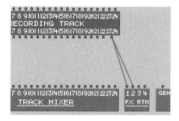

Fig.: Bouncing FX Return signals

For reasons of efficiency, transmitting such signals to external devices is recommended via →Direct Paths. Assign the Return signal to any DIR pair and route the pair to the desired outputs (→Output Assign). Connect the external device to those outputs and then continue work.

FX RTN

Short for →FX Return Channel on a mixer that allows →Effect Return signals to be processed.

FX Rtn SOLO ENABLE

Only the VS-2000 provides an F. function that automatically soloes the FX Return channels when a mixer channel is →Soloed (→AFL). This allows that signal to be heard in isolation with →Effects applied (→FX Return Channel).

*UTILITY → PLAY/REC → Param1 →
"FX Rtn Solo Enable"= ON*

G

Gain

An →Equalizer's Gain (G) parameters allows the boosting (increase) or cutting (reduce) of the level of the associated filter band. The setting range is usually ±15dB (→Massenburg EQ ±24dB).

Gate

A mixer channel's →Expander can be used as gate effect by setting →Ratio= ∞.

Gate Reverb

Apart from producing the famous "Phil Collins" snare/tom sound, this →Effect can also be used for reverse reverb effects (reverb that goes backwards). A G. stops abruptly, which is especially effective for experimental sounds. A reverse reverb effect, on the other hand, rises slowly, and then stops abruptly.

Gate/Reverse Reverb	
Effect patch:	P074~P077
Connection:	Insert
Application:	Special effect

Editing

The algorithm can be specified with "Mode": "Normal", "R->L" and "L->R" represent gate effects, R->L / L->R has an interesting stereo effect. Revers1 and 2 generate reverb effects that seem to run backwards, which is due to a rising and falling level envelope.

Generator/Oscillator

The VS-24xx's generator can output →Pink Noise, →White Noise, a tunable →Sine Wave, and →Metronome sounds.

Application

- Test tones for internal and external use
- Recording the drum computer's signals to a track
- Checking a room's resonances via Pink Noise and →Analyzer

Only the LCD contains additional level meters for AUX Send and Direct Path assignments:

VGA:	Utility → Gen/Osc
VS-24xx:	Utility → Osc/Anlyz

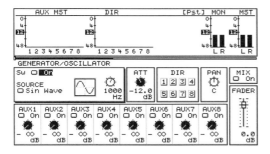

Fig.: Signal tone generator in the LCD

After activating the generator, use "Source" to select the desired signal source, and route it to the →Mix Bus (Mix= On), an →AUX Bus or a →Direct Path. Only the LCD allows the →Level Meters for AUX or the Outputs to be called up in order to check the proper levels.

Tip

The oscillator can also be used to check the routings of analog inputs. To do so, route the oscillator to a Direct Path, and from there to an Output socket. Connect a cable between the AUX OUT socket and the desired input, and start checking.

Rhythm Computer (VS-24xx)
To record the internally generated drum, go to the →Metronome and set "Tone Typ" to "DRUM". In the Generator, select "Metronome". The routing required for recording the signal is shown below.

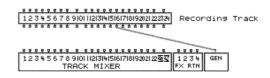

Fig.: Routing the generator to a track

Get Now

In →Numerical Editing, parameters that require a time position, take advantage of the F2 (Get-Now) function. This copies the timeline's "current time" into the selected field so that it doesn't have to be set manually. →Go To, on the other hand, allows the position shown in the current field to be located.

Global Parameter

The Global pages 1 and 2 of the →UTILTY Menu contain parameters usually needing to be set only once, such as →Input Peak Level, →CD Digital Rec, →Foot Switch, →Previous/Next Switch, etc. On the VS-2000, the Global parameters are part of the →System Parameter.

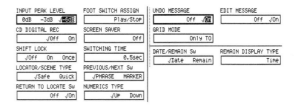

Fig.: Global Parameter 1 and 2 pages

Go To

For →Numerical Editing based on time positions (e.g. Region and Phrase Edit), F3 (Go To) can be used to jump to the position indicated by the input field. This can be used to double-check the settings. Unlike G., →Get Now copies the current position to the selected field.

GPI

The "General Purpose Interface" standard allows the VS to receive →Foot Switch Start and Stop commands from a special video device.

GR

The →Dynamics pages of the mixer channels contain a GR meter that displays the gain reduction achieved with the current settings.

Gradation

The G. function generates new →Automix settings between two instances that use different values. This can be used to smooth transitions, create fade ins/outs, etc. (See →Gradation (p. 23) in the Automix chapter).

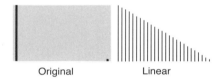

Fig.: Gradation of Automix data (example)

Graphic EQ
→Graphic Equalizer

Graphic Equalizer

A graphic equalizer divides the audible frequency spectrum (±20Hz~20kHz) into equal ranges whose level can be set using separate controls. An →Octave band EQ, for instance (see →Effect Patch P212~214), provides 10 bands whose center frequency doubles at each step: 32.2, 62.5, 125, 250, 500, 1k, 2k, 4k, 8k, 16k. A third-band EQ, on the other hand, has 31 narrow bands (distributed in 1/3-octave steps). The fixed center frequencies of a G. are much clearer and easier to work with than a →Parametric Equalizer.

Graphic Equalizer	
Effect patch:	P212~P214
Connection:	Insert
Application:	Bus

To eliminate any "muddiness" in mixes, the 250Hz band of such signals can be cut. A G. is usually only used for busses and compensating for room resonance.

Grid

In the VGA, "Grid" simplifies editing operations like copy, move, insert, etc. (→Region Edit, →Phrase Edit, →Automix) by enabling work with a magnetic grid. This only works if the Project tempo (the internal →Metronome or an external synchronized →MIDI Sequencer) was used while recording. Grid quantizes →Edit Points by moving them to the nearest time location of the adjustable grid. For instance: Grid= MEAS means that moving an item to 002–03 (third beat of measure 2), it will positioned automatically to 003–01 (first beat of bar 3).

Application

- Precise arrangement of individual Phrases or Regions of one or several tracks.
- Trimming Phrases in bar or beat steps for an intended arrangement (→Phrase Trim).

Activating And Setting The Grid Step Size

Right-click the →Playlist on the VGA screen to open the edit pop-up menu. Select "Grid" in the bottom row to call up the Grid window. Here the step size can be selected between "Measure" (bar) and "1/32".

Alternatively, this window can be selected by clicking the "GRID" field in the →Measure Bar (upper right corner).

VGA:	Click "Off" in the Grid field and select a value
Mouse:	Rightclick in the Playlist and select a value

VGA.: Activating Grid in the VGA

1. Before activating the Grid, trim the →Phrase beginning so that it coincides with the first notes needed (→Phrase Trim). →Preview, →Scrub and →Waveform will help to perform this task more quickly.
2. Activating the grid.
3. Drag the Phrase to one of the visible raster lines.

After performing the steps above, the Phrases and Regions can only be moved to other grid lines – positions between two vertical lines are impossible. If a Phrase's start point was not on a grid line when the function was activated, the corresponding offset (with respect to the grid line) is maintained when the phrase is moved or copied. In this way, even graphic →Mouse Editing can be performed with an anchor point (→FROM).

Tip
To move a Phrase or Region to a place not supported by the grid, hold down CLEAR while dragging it. Doing so temporarily bypasses the quantize setting.

Quantizing Only TO
Some applications may prove more efficient if only the TO position is quantized. This allows a Region to be defined without any restrictions while the destination is tied to a rhythmic grid:

Global → Param2 → Grid Mode

Ground Loop
→Hum (p. 94)

Group
A fader group allows the level of several mixer channels to be set by moving just one fader (see Subgroup). The VS provides 12 fader groups that can be used universally.

Application
- Maintaining the balance of backing vocal tracks while leveling them.
- Changing the overall level of the drum or percussion section (also possible on the Input Mixer level for a drum machine).
- Changing the levels of all effect return channels with just one fader.

Activation
- Set all faders of the group to establish the right balance.
- Use →CH EDIT "GROUP" to specify a group number for the channels.

The →Multi Channel View shows all channels of the current mixer section, which may be easier. Groups can span different sections, e.g. INPUT channels 3~5, TRACK channels 12~18, and FX Return channels 1 and 3.
Select "Off" to remove unwanted mixer channels from a group.

Making Corrections
If the level of just one channel assigned to a group needs to be corrected, hold down CLEAR while moving its fader.

Guitar Amp Modeling
The G. →Effect uses →COSM modeling to simulate the sonic behavior of a guitar amplifier, providing a variety of readily usable guitar tones. There are parameters for both the preamp and the simulated speaker box, which adds to the effect's realism. Authentic guitar amp sounds can be achieved by using G. and the →Guitar Hi-Z jack. This saves the time of with a real recorded guitar amp.

Guitar Amp Modeling	
Effect patch: [GA]	P160~168
Connection:	Mono insert
Application:	Standard (guitar)

Given that the signal level is often boosted heavily by a guitar preamp, a →Noise Suppressor is located before the preamp stage to block unwanted noises. Use the "AmpType" param-

Guitar Hi-Z

eter to select the amplifier sound (Fender, Marshall, Boogie, Soldano, Peavy, etc.), and the knob icons to set the tone (they act in the same way like the originals). The "Speaker" field allows the desired guitar speaker sound to be selected (It is even possible to select combinations that don't exist in the real world). Use "MicSetting" to specify the position of the virtual microphone used to capture the sound and add the direct signal ("DirLvl" 1= close, 3= much ambience).

Tip
For special sound designs, the coloration of the speaker models can be used without the preamp by switching off the remaining blocks.

Alternatives:
- →McDSP ChromeToneAmp (p. 119)
- →GuitarMulti (p. 88)

Guitar Hi-Z

This Hi-Z input allows electric guitars and basses to be connected. Due to their high impedance, they cannot be connected to the VS's "regular" inputs. When the GUITAR HI-Z function is switched on, microphone input 8* is no longer available.

Tip
The Guitar button could be used as an on/off switch for input 8*, making it the ideal candidate for →Talkback applications.

VS-2480: Input 16

GuitarMulti

The "Guitar Mult 1~3" →Effects are simulations of various guitar multi-effects processors with the addition of a speaker simulation. The sheer number of effect blocks allows creating a wide variety of sounds. The only difference between the patches is the preamp function (second block).

The available blocks are:
→Compressor, preamp, →Noise Suppressor, Auto-Wah effect, speaker simulation,

Guitar Multi 1, 2 and 3

Effect patch: [GT]	P150~159
Connection:	Mono insert
Application:	Standard for guitar

→Flanger and →Delay.

Mlt	Preamp	Description
1	Heavy Metal	Extreme overdrive, numerous harmonics
2	Distortion	Vintage sound, fewer harmonics
3	Overdrive	Treble boost, the mid and high ranges are emphasized

Alternatives:
- →McDSP ChromeToneAmp (p. 119)
- →Guitar Amp Modeling (p. 87)

H

H.Position
In the →VGA Monitor H. allows a horizontal Shift of the image. Use this function, if the image is not located in the center of the seen. The setting range is –5~+5.

Hard Disk
The H. (hard drive) is a typical digital data storage system. It contains several rotating disks with a magnetic coating. These are accessed by the machine's so-called read and write heads. Because of the high speed at which the disks rotate, the heads never actually touch the disks. Instead, they float on an extremely thin cushion of air (a few micro-millimeters), which means that the machine has a certain mechanical sensitivity (wait until the disk has come to a halt before transporting it (→Safety in Operation)).

Before using the hard drive for the first time, or to carry out a →Defragmentation, it must be →Formatted and →Partitioned. Data is recorded onto the hard drive as files. Because of the usually considerable size of the data, it is stored in logical, consecutive →Sectors.

For reasons of quicker and safer data retrievals in the VS, the H. (40~80GB) is subdivided into several →Partitions of 10GB each. This has an important effect on the available →Recording Time.

Tip
The →Drive Check operation enables a check of how the H is functioning. After →Format the H, the drive's state can also be examined using →Surface Scan.

Hard Disk Recorder
An H. records audio signals as data which is stored on the →Hard Disk. If such a device additionally provides all components needed for studio productions, it is often called a "→DAW".

Harmony

1. Overview
The Harmony function is exclusive to the VS-2000. It generates choir voices from one single recorded vocal or instrumental track. The notes to be added can be created using the VS's buttons, a MIDI keyboard or with the internal harmony sequencer. Harmony is based on the →Voice Transformer algorithm and allows the →Formants, Pan, Portamento and Vibrato of the generated voice to be changed. For the production of harmony voices, the processing power of one or more VS8F-2 →Effect Boards is required. These boards are then no longer available for further effects. The VS8F-3 →Plug-in boards cannot be used for HARMONY voice generation. Each FX board can generate two voices. Three-part harmonies therefore require the installation of an additional board. Because the harmony parts are generated in an effect processor, they can be processed through the corresponding →Effect Return channels.

2 Setting Up
Despite the impressive number of adjustable harmonizer parameters, setting up is surprisingly simple. The parameters can be saved in →Scenes and retrieved whenever necessary. This should allow the function to be used creatively. Only the Poly/Mono switch and the Fine Tune settings are not saved by the Scenes or Harmony Patches. These have to be changed manually.

2.1. Startup Page
Harmony Assign

The main harmony settings are carried out on the "HARMONY ASSIGN" page. To select, press the HARMONY button on the VS, or select it in the VGA's effect menu.

Harmony

HARMONY ASSIGN

USE Effect BOARD:	Specify the number of parts*, Harmony" is activated. Deactivate: Off (see fig.)
HARMONY SOURCE:	Source signal for the harmony voices (see fig.).

   * *When shipped, the VS-2000 includes FX1 and FX2, which can be used for two voices (parts).*

2.2 Algorithm Page

Pressing F1 (Edit) calls up the "HARMONY ALGORITHM" page. The settings for this page are, for the time being, irrelevant and can be set at a later stage.

The "HARMONY ALGORITHM" Page

TR x SEND (→Pre/Post)	"Pre" allows the source signal to be muted. In this way, the harmony parts only can be heard.
HARMONY sequencer	Switch on the sequencer to record and playback notes.
MIDI	For inputting notes via a MIDI keyboard.

2.3 Changing the tone characteristics

All settings saved to Patches can be changed on two edit pages called "Modul Common" and "Modul Part".
The COMMON parameters apply predominantly to all parts.

COMMON Edit

CONTROLLER	Poly: For Polyphonic parts Mono: Monophonic parts. Forces a constant sound, which is necessary for the Portamento.
FINE TUNE	Overall tuning between 435Hz and 445Hz.
BEND RANGE	For incoming MIDI Pitch Bend messages. This sets the maximal interval in semi-tone steps.

PORTAMENTO Sw: Time:	Glides from note to note. Switch on/off (see Controller) Time taken before the next note is reached.
VIBRATO Sw: Rate: Sw: Rate: Detune: Depth: Delay:	Adds a frequency modulation Switches vibrato on/off Vibrato speed [Hz] Switches vibrato on/off Vibrato speed [Hz] Sets the amount of detuning Sets the strength of vibrato Delays the vibrato effect

The Part parameters allow the tone color of each voice to be changed independently. This can be used to establish a clear distinction between the bass, tenor and soprano voices, to set the balance and their stereo distribution.

PART Edit

FX RTN LEVEL	→Pre/Post setting for level measurement of the →FX Return Channels*.
FORMANT	Sets the voice timbre (e.g male/female, →Formants)
PAN	Specifies the location of the selected voice within the stereo image.
LEVEL	Sets the voice's volume.
F2 TX Prm	Records the parameter value with the →Harmony Sequencer.
F3 All	Records all parameter values with the →Harmony Sequencer.
F4 PATCH	Calls up the list of Preset and User Patches.
F5 SAVE	"SAVE" page where changes to the Common and Part parameters can be stored.

   * *If the harmony voices are inaudible, use PRE to check whether the signal source is received. Next, set the FX Return channel's fader to "0dB".*

Harmony

3. Generating Harmony Voices

3.1 Using the Track Status buttons

The two default settings of the "Harmony Assign" page already allow playing a second voice to a mono vocal track in real time. While the Harmony function is active, the TRACK-STATUS buttons act as a simple keyboard which can be used to play a melody. Use buttons 5~11 to play a simple melody in C major. It is even possible to generate two- or three-part* harmonies, providing a certain dexterity when inputting. The octave of the generated voice can be determined with the 1 key (to lower by up to two octaves) and 2 (up one octave) buttons. The legato function assigned to button 4 is used to generate legato-style melodies. Only in MONO mode, the PORTAMENT function is assigned to button 5 (see the tip below). The remaining buttons are used to enter chords.

Tip:

Consider working in Mono mode whenever a one voice melody is needed. This mode provides the best sound quality and a preset Pan assignment, which avoids additional editing because the voice is always output by the same part. MONO is also a prerequisite for the Portamento function. Used in small doses, it simulates the natural behavior of the human voice.

Edit → Module → "Common" page → Controller = Mono

The FX boards installed at the factory allow two voices to be generated. Additional boards will add two voices each.

3.2 Using A MIDI Keyboard

It is easy to configure the settings for playing harmony tones via an external MIDI keyboard. Connect the VS-2000's MIDI IN socket to the keyboard's MIDI OUT port, jump to the "ALGORITHM" page and select the MIDI channel in the "MIDI" section. The channel must be the same as the one assigned to the MIDI keyboard ("xx" refers to the keyboard's MIDI channel).

F1 (Edit) → "Algorithm" page → MIDI= xx*

Tip

- If chords are outputted, even though a one line melody on the keyboard is played:

    "Realtime" page→ Input Mode → "Note"

- Depending on the number of additional →Effect Boards, chords can also be played on the keyboard (Legato= Off).

- By default, incoming Pitch Bend data with the maximum or minimum value will transpose the harmony voice up or down by one tone. This interval can be changed via the "Bend Range" parameter on the "Common Edit" page.

- Though the harmony source is set to be a recorded track, the effect can also be added to a live, sung vocal part:
    To do so, route the input signal to the track defined as "Harmony Source" and engage its recording function.

3.3 Entering Chords via VS's Buttons

Even less than accomplished keyboard players can enter chords without difficulty. The most commonly used chords are assigned to TRACK STATUS buttons 12~17, with buttons 1~11 allowing the fundamental/root to be specified.

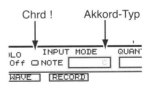

Simple applications and chord changes, which aren't too fast, can even be played in real time with the VS's buttons. The approach described below is rather more analytical and therefore chiefly for step-recording applications (→Harmony Sequencer). After activating the Harmony function, set the INPUT MODE parameter on the "Realtime Record" page to "Chrd":

F1 (Edit) → F3 (Realt) → Input Mode = "Chrd"

Harmony Sequencer

With the default setting of "Major", the buttons 1~11 already allow the corresponding major chords to be played. Additional chord types can be obtained by setting the mode (button 13), type (buttons 14~17), b/# (button 12) and octave (buttons 1 and 2). The chord is displayed by the "MODE" field in relation to C major. The following table shows which buttons produce which chord types:

Type	Button	Color	Mode
Major	13	Red	C
Minor	13	Green	Cm
7th	14	Red	C7
Major	14	flashes red	CM7
6th	14	Green	C6
9th	14	Orange	C9
Augm.	15/16	Green	CAug
4th	15/16	Red	Csus4
Dim.	17/18	Red	Cdim
ST lower	12	Flashes green	b *
ST higher	12	Flashes red	# *

ST= semi-tone

* *Displayed above the virtual keyboard*

Patch

Patches not only contain sound-related part parameters (for the various voices), but also predominant effects like vibrato, portamento and overall tuning. In addition to the 50 Presets, some of which sound amazingly authentic when used in the right register (P06, P09, P13, P19, P44 [activate the Mono setting], P46, etc.), 50 User patches are available to store individually created settings. The "Patch" page can be accessed both from the opening Harmony page ("Assign") and the two edit pages.

Harmony Sequencer

The H. can be used to record the desired →Harmony notes in real time or in step mode. In addition, notes can also be edited using "Micro Edit". Select the "Harmony Algorithm View" page to record the desired note events.

Real-Time Recording

F3 (Realtime) opens the "Realtime" menu where the quantize grid and assign note, Pitch Bend and CC messages (below "Target") can be set. Start recording with F2 and play the notes following the description for →Harmony (p. 89).

Step Recording

For projects, recorded with click, pressing F4 (StpEdt) enables a step by step entering of the harmony notes. Start by setting the STEP size for the notes being entered. The note duration can, of course, correspond to the step size, but it can also be based on a different value. Each time a note or chord is played, it is recorded and the step counter (and the Project's timeline) advances by one unit. Rests can be inserted between steps and backstep used to return to a previous step.

Headphones

→PHONES (p. 146)

Headphone Mix

An H. is a special mix prepared for musicians during a recording session. As it is sent to the →Studio Monitors independently of the main Mix, it can be used to cater to the musicians' preferences. In sophisticated setups, each band member or session player gets their own H. The number of headphone lines used depends on the number of →AUX Busses available on the mixing console. Consider using the VS's TRACK MIXER for outputting the AUX signals, as that is the only way of ensuring that the signals remain audible during playback and during →Punch In/Out recording. In most instances, two AUX busses are linked to a stereo pair. The signals to be transmitted should be sourced before the faders (→Pre/Post) to ensure that previously recorded signals are always audible in the cans – irrespective of how the faders are set for the main mix. Preparing a →Talkback microphone for communication with the musicians is a possibility.

Effects For The Headphone Mixes
Example: Using Effect 1 to add a reverb effect to the headphone mix of AUX busses 7/8.

1. Assign the reverb effect to the mixer channels (→Send/Return Effect).
2. Select →FX Return Channel 1 via →CH EDIT.
3. Set the AUX7/8's Send knob as desired.
4. If necessary, use the MIX switch to remove the effect from the mix bus. MUTE removes the effect signal from the AUX bus.

Headroom
In digital audio systems, a headroom of 12 dB must be provided to avoid intolerable distortions (→Clipping). H. stands for the difference between the average operating level and the maximum level that the VS can let pass without distortion (→Level Setting While Recording). Due to the converters' 24-bit conversions, such signals are then coded with 22 bits, which represents a →Dynamic range of ±130dB. In contrast, with this 12 dB headroom, 16-bit converters (CD quality) only offer a dynamic range of a mere 80dB.

High
This term refers to the →Equalizer band that allows high frequencies to be set (usually a shelving filter).

High-Mid
On a VS, the upper peaking filter of an →Equalizer section is called an H. The peaking filter's center frequency can be set to any value of the supported frequency range.

High-Pass
Only the VS-2480's →Equalizer provides an additional multi-purpose filter that can be set to "HPF" (high-pass). It allows all frequencies at and above the selected value to pass, while lower frequencies are attenuated at 12dB/octave. The "dB/octave" refers to the filter's slope. If an HPF's frequency is set to 1kHz, a 6dB/octave slope means that the 500Hz frequency is attenuated by 6dB, 250Hz by 12dB, 125Hz by 18dB, etc. (→High- and Low-pass filters of the VS-2480).

History
Within the →Name Entry function, the "History" function allows the recall of any name entered since switching on the VS. If naming tracks, channels, etc., has become a habit, this function may help save a lot of time.

HOME
The VS's "Home" page is the most important display page. It displays important information about the Project, the current timecode location and the editing status. The "Home" page can be recalled from almost any edit or system operation:

VGA: Click HOME (menu bar, right)
VS: HOME button
VS + mouse: Click HOME (upper left corner)

The "Home" page contains the following elements (to name a few):

- Track display with track names
  →Playlist (p. 158)
- Numeric representation of the current timecode location, with bar display and markers
  →Time Display (p. 224)
- Level meters
  →Level Meter (p. 103)
- →V-Track (p. 233)
- →Scene (p. 200)
- →Locator (p. 106)

Hum

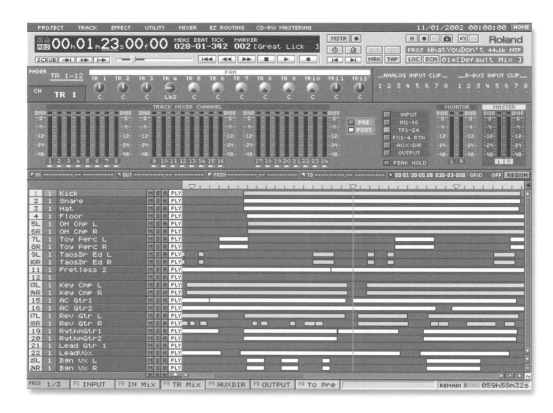

Fig.: "Home" page on the VGA screen

The →Information Display (usually the LCD) can display the Home page irrespective of the page currently shown on the VGA screen.

Hum

Differences in electric potential, usually due to repeated grounding of power and signal lines, may induce an unpleasant hum. On 50Hz power lines, its frequency usually corresponds to 50 or 100Hz. The following should enable it to be suppressed:

1. Use the same power phase (connect all devices to the same power outlet)
2. Transmit signals in the digital domain
3. Use a →D.I. Box

Insert the →Hum Canceler effect

Hum Canceler

A line hum (ground loop) can be removed using the →Effect H. It suppresses the line frequency and its harmonics when an audio system has multiple paths and path lengths to ground. Line hum reduction affects the overall audio quality minimally and it should only be used as a last resort (→Hum). →Insert the effect (as "InsL" or "InsR" for mono tracks) and set the frequency to 50 or 60 Hz, depending on the country. Lowering the THRESH level also allows unpleasant disturbances to be suppressed. For high-frequency noise, this parameter can be set to "6kHz". The canceler is followed by a →Noise Suppressor that mutes conventional low-level noises.

Hum Canceler	
Effect patch:	P032
Connection:	Insert
Application:	Troubleshooter

I

Icon
Depending on its function or position, the mouse →Cursor may change shapes. This is usually done to expedite →Editing operations.

ID ChV
Short for "Information Display, Channel View". With the Connection of a VGA screen two displays are then available: The so-called →Operation Display (usually the VGA screen), where operations are performed, and the seemingly unnecessary →Information Display (usually the LCD) called "ID". When using the VGA screen, the I. can additionally show the →HOME or →CH EDIT page, the →Waveform, or the →Analyzer (not for the VS-2000).

Page 3/3 → F2 (ID ChV)

The page to be shown in the Information Display can be locked with →IDHold.

IDAnlr
Short for "*Information Display*, →Analyzer", the I. is used to independently display the Analyzer in the LCD (→ID ChV).

IDE Drive
On the VS, each →Hard Disk →Partition shown in the →Project List is called an "IDE drive". It refers to the use of a selectable IDE ("Integrated Device and Electronics") hard disk.

Tip
Using an "IDE to SCSI converter" allows cheaper) external IDE drives to be connected to the VS-2480.

IDHold
When working with an additional VGA screen, a page selected with I. can be held in the LCD and hence doesn't disappear when a different page is selected on the VGA screen (→ID ChV (p. 95)).

IDHome
Short for "→Information Display *Home*". When using a VGA screen, I. can be used to independently recall the →HOME page in the LCD.

IDWave
Short for "Information Display, →Waveform" (see "ID ChV").

IK Multimedia
→T-Racks Mastering (p. 220)

Image File
→Disc Image File (p. 57)

Import
Projects and Songs created on other VS models can be loaded onto the VS for additional recording or editing sessions. If the →Project Recover operation triggers a "Cannot Execute! Other VS Project found!" message, the Project was most likely created using a different model. In this case it is only possible to load it using the "Import" operation.

Project → List → Page 3/4 → F3 (Import)

This operation is similar to "Project Recover". The current model must have at least the same number of tracks and →Record Modes.

Import Track
→Track Import (p. 228)

Import Wave
The →Wave Import (p. 242) function enables importing WAV-format audio data created on other systems via CD-R.

IN

IN
The IN and OUT →Edit Points specify the beginning and end of the region to be edited (→Region Edit or →Automix editing).

IN F/P
The "→Input Mixer, Fader/Panorama" page on an external VGA screen displays the input channels along with certain parameters and the →Aux Master faders. The →Dynamics section, →EQ, →Solo, →Mute, and →Automix can be activated, and the AUX and FX Master levels can be set. The Master Send settings and →Direct Paths are also shown.

Home → Page 2/3 → F2 IN F/P

Even when shown in the LCD, level and Pan settings can be controlled and changed here with the mouse.

IN Mix
The →Function Buttons can be used to call up the desired →Level Meters. I. refers to the →Input Mixer level meters.

PAGE1/3 → F2 (IN Mix).

Info
I. displays additional information about the size, creation date, name, etc., on the →Phrase New and →Take Manager pages.

Information Display
When working with an external →VGA Monitor, the screen not used for editing (usually the VS's LCD) is referred to as ID. It is possible to alternate between the Information and →Operation Displays by pressing:

VGA:	Utility → System Parameter2 → VGA Out
VS-2480:	SHIFT + UTILITY
VS-2400:	SHIFT + MASTER EDIT
VS-2000:	HOME + F6

The I. can be used to display additional information, such as →Channel View, →Waveform or →Analyzer. To achieve this, select page 3/3 via PAGE in the playlist display and assign the desired display function to it. The page to be shown in the Information Display can be locked with →IDHold.

Page 3/3 → F2 (ID ChV)

Init PB, IniTrA
This operation allows the default settings for the patchbay and the →EZ Routing function's track assignments to be called up.

Initialize Harddisk
→Format (p. 80)

Initialize Mixer
The mixer parameters can be reset to the default values (→Prm Init (p. 165)).

Utility → Prm Init

INPUT
Refers to the →INPUT Level Meters (input section). The →Function Buttons can be used to call up the desired →Level Meters.

PAGE1/3 → F1 (INPUT)

Input Clip
I. refers to the VS's peak indicators in the analog input section. They light to indicate that the →Headroom has been exceeded well before the signal starts clipping. Please note that only the →INPUT Level Meters can provide the most exact information about input levels. While recording several signals simultaneously, certain clip indications may be overlooked, which is why it is usually a good idea to lower the →Sensitivity Knobs of channels with a lit I. The Input Clip indications are displayed both in the VS's LCD and on an

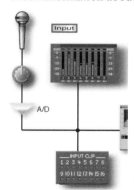

external VGA screen. The →Input Peak Level parameter allows the threshold to be set to –6, –3 or 0dB. The default setting is "–6dB" (which is restored by the →Initialize Mixer operation).

Input Impedance

When working with analog inputs, the →Microphone Preamp constitutes an impedance (→Analog Inputs). The signal source's impedance should be higher than that of the VS input, as this eliminates coloration caused by impedance mismatches. The VS also provides a Hi-Z input for electric guitars, etc. (→Guitar Hi-Z).

Input	Impedance
1~8 (16)	10kΩ
Guitar HI-Z	1MΩ

INPUT Level Meter

The INPUT level meters indicate the levels measured right behind the A/D converter – i.e. before the input channel. They are used to control the levels of analog and digital input signals, and are sourced from the same measuring point as the →Input Clips (→Level Setting in the Input Mixer (p. 177)).

In order to produce a large →Dynamic range and an optimal →Resolution, the signal levels should be set as high as possible. There should only be occasional →Peaks that exceed the -6dB →Input Peak Level, 0dB (OVER), must be avoided at all times. The following illustration shows the location of the measurement points on an →Input Mixer channel:

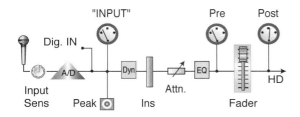

Fig. Level indications of the INPUT MIXER section

Input Mixer

Among the various →Mixers on a VS, the INPUT MIXER is used to process input signals which are being fed into the analog and digital inputs via the →Patchbay. While →Playing back, it routes the input signals directly to the →Mix Bus. This has the advantage that it can be used alongside the tracks, and hence makes it possible to add MIDI devices, CD sources, FX Returns of external processors, etc., in realtime. During →Recording, the I. channels transmit the received signals to the tracks activated for recording. The faders should usually be set to 0dB.

Fig.: Location of the Channel Strips

VS-2000

The eight analog and two digital inputs are hard-wired to the corresponding INPUT MIXER channels. There are no physical faders or controls available. Using the (round) INPUT section buttons the desired →Channel Strip can be selected. After moving the cursor to the fader field (… dB), the setting can be changed via the TIME/VALUE dial.

Input Peak Level

The I. parameter is used to specify the input level at which the peak indicators should light (→Input Clip (p. 96)). This level is indicated by means of a horizontal line in the LCD. The (sensible) default setting is "–6dB". It is also possible to select "–3dB" or "0dB (Clip)"(→Recording). In the VS-2000, this parameter is to be found in the System menu.

Utility ➔ *Global1* ➔ *Input Peak Level*

Inputs

Inputs

The physical inputs are used to feed analog or digital signals into the VS system via the corresponding input sockets (→Analog Inputs, →Digital Inputs and Outputs). The VS considers them building blocks of the →Patchbay, allowing them to be routed to the desired →Input Mixer channels using virtual patch cables.

Insert

"Insert" refers to an effect connection whereby a signal path is interrupted and the →Effect is inserted. The processed signal returns to the same point from which it left in the particular channel without affecting any other channels or buses.

A →Send/Return Effect connection (for reverb, chorus, delay etc.), on the other hand, works with a copy of the original signal.

The connection system means that an insert effect can only be used for one channel (or a stereo channel) at a time. So-called dual mono effects can still be set and used separately (for the left and right channels). Even →External Effect Processors can be included as insert effects on the VS.

Application

Use as additional →Compressor, →Equalizer; effects for →Electric Guitar; →Vocoder, →Mic-Modeling, etc.

Connection

Insert effects can be assigned both to INPUT MIXER and TRACK MIXER channels. The insertion point is located between the →Dynamics and →EQ section of the →Channel Strips, but before any Send line.

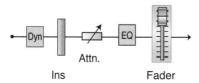

Fig.: Insertion point in a channel strip

Use →CH EDIT to select the desired channel.

VGA:
Click the "Off" slot in the "Insert" field to establish the connection.

LCD:
Select "CH EDIT", move the cursor to "Fx Insert", and confirm with [YES]. The channel's "Insert" page is displayed. Move the cursor to a slot and turn the dial once to the right.

VS-2400:
The FX INS key in the CH PARAMETER section calls up the "Insert" page for the last mixer channel selected.

Insert Modes

Insert effects can be connected in several ways; by rotating the dial several steps towards the right, or by clicking several times. In most instances, the mono connection will be used. By merely deactivating an insert connection, the other possible varieties for the independent usage of Dual-Mono effects become visible. The VS's LCD displays the selected insertion mode.

1. "Ins" For Stereo Channels
For stereo channels and stereo effects, this is the only form of connection available. Be aware that a mono effect will position both signals at the center of the stereo image.

2. "Ins" For Mono Channels
For mono effects (like →GuitarMulti) in a single channel, "Ins" is the right connection. Wrong use of a stereo or dual-mono effect (i.e. different settings) may lead to →Phase Cancellations, which translate into a slightly nasal and hollow sound. To avoid this problem, select one of the following modes:

3. "InsL/R" For Mono Channels
Dual-mono effects provide two separate effect blocks that can be inserted into different mono channels and hence used as completely independent effects. The →MicModeling effect of blocks A (InsL) and B (InsR), for instance, could be used to create two different microphone simulations.

4. "InsS" For Mono Channels
In a Dual-Mono effect, the signal runs through blocks "A" and "B" consecutively (serial connections). This represents a serial connection of blocks "A" and "B" for dual-mono effects. The signal is therefore processed by two effects.

Tip
A dual-mono effect connected in a series turns a 5-band EQ into a 10-band EQ (see also →Massenburg EQ (p. 110)).

Insert Send and Insert Return
When inserting an effect via the LCD, the terms "Send" and "Return" are prominent, which could be confused with a →Send/Return Effect. These seldom used parameters make additional level controls available, which in turn allow the level before and after the insert effect to be regulated. Some guitar effects may generate a desirable coloration in response to different input levels. Effect chains consisting of several effects connected in sequence, may also benefit from different input and output levels.

On the VGA screen, the levels can be set using the two bars on the right next to the yellow effect symbol.

Master Section
Effects inserted into the →Master Section process the entire mix bus. This is the same as using the →Mastering Room's inserts. A →Master Insert with External Effect (p. 116) can also be used via a special routing.

VGA:	Mixer → Master Edit	
VS:	SHIFT + MASTER	
VS-2400:	MASTER EDIT	

INT

I. →Synchronization (p. 217)
In the normal operation mode, "INTERN", the tempo of the Project's reproduction is governed by the VS itself. Outgoing synchronization data (→MTC, →MIDI Clock) for external slaves is based on this tempo information (see EXT).

II. Master Clock
→Master Clock (p. 111)
In the normal mode, the VS's clock generator provides the →Sample Rate itself.

Interface
A connector and protocol that allows hooking up devices using different digital audio formats are called I. This requires the necessary hardware and software. The VS provides the following interfaces for digital audio signals: →S/PDIF (and →AES/EBU via →AE-7000) and →R-Bus.

Isolator, 3-Band
Unlike conventional →Equalizer (p. 68)s, the "Isolator" →Effect (p. 62) suppresses the frequencies of the selected band almost completely. This is similar to working with a crossover unit. The frequency spectrum is divided into Low, Mid and High bands whose levels can be reduced by up to 60dB (40dB for the Mid band). The cross-over frequencies are located at 500Hz and 3kHz. An I. can be used for special effects (removing the bass and kick drum from a mix), or to eliminate, undesirable, high-level noise.

3 Band Isolator

Effect patch:	P028
Connection:	Insert
Application:	Experimental

J

Jogwheel
Only the VS-2480 provides a jogwheel for fast positioning. This function, known from most video recorders, allows the speed to be controlled by varying the degree to which the wheel is turned left or right.

Joystick

Joystick
The →VE-7000 channel strip is configured for →Surround panning with the J.

Jump (1)
Use JUMP to locate a desired time position within a project. The time value can either be entered using the cursor buttons and the dial, or with the →NUMERICS function (numeric key pad).

    VS-2400:   SHIFT + PAGE
    VS-2480:   JUMP

Application: For direct positioning of an exact time position, similar to a →Locator. This avoids, for example within →Automix, any unwanted automated settings.

Jump (2)
As the VS-2000 is not equipped with motorized faders, capturing a stored level value can be specified after recalling a →Scene.

The "Jump" setting means that the stored value is used as soon as the fader is moved– provided the current fader position is different from the stored value. This setting is used by default, as any change is immediately reflected by the level of the channel being controlled.

Null
The "Null" setting means that the stored value is only used when the fader reaches the physical position corresponding to the value. This means that the fader appears to do nothing for a certain time.

Use the →Fader Match (VS-2000 only) parameter to select either "Jump" or "Null":

    Utility → System → Fader Match

K

Key In
In most instances, the signal itself is analyzed to establish the required gain reduction by a →Dynamics processor. In exceptional cases, this doesn't lead to the desired result, particularly for signals with extremely small level differences between useful signals and unwanted noise disturbance signals. Here the signal of another mixer channel can be called upon to aid the detection. This mode is called "Side Chain" or "Key In".

Track Copies With Biased EQ Settings
If the dynamic differences between the wanted and the unwanted signals are too small, consider copying the track to be used as Key In for the →Expander and selecting extreme EQ settings for it. For example: Cleaning up a snare track from HH and BD crosstalk by using the copied Key In track with extreme cuts for the high and low ends.

Tip:
To optimize the expander's response, consider shifting the copy 1~2 frames towards the Project's beginning ("Look Ahead" approach).

Triggering An External Signal
The low-frequency content of a kick drum can be boosted by means of a sine wave (50~80Hz), while the snare carpet part of a snare drum can be emphasized using white noise. Both additional signals can be provided by the tone generator (→Generator/Oscillator). Select →EZ Routing, route the generator to a channel and record its signal. For the corresponding mixer channel, activate the →Expander and select the BD or SD channel as "Key In" signal to ensure that the gate opens and closes in the right places. The following settings are recommended: Ratio −∞; Threshold ±10dB below the generator's level; Attack = 5~10 ms; Release for the sine wave: at least 200ms.

Ducking
"Ducking" refers to lowering the level of the signal being controlled in relation to the volume of another signal. This is often used for voice-overs: when the commentator starts talking, the background music automatically becomes softer. For this effect, set the

→Compressor to use a 1 : ∞ Ratio, and a Threshold value of –24dB. Switch off Auto Gain and select a long Release time (±4s).

Keyboard PS/2

Particularly →Name Entry, but also starting playback, and selecting dedicated menus can be performed advantageously using an optional PS/2 keyboard (the VS-2000 needs to be fitted with an optional VS-20 VGA board for this to work). The following operations can be performed using such a keyboard:

Table: Operation via a PC keyboard

Keyboard	VS button	Menu/function
F1~F6	F1~F6	Different assignments
Esc/Pos1	HOME	Back to Playlist
F7	EZ Routing	Routing View
Shift + F7	SHIFT + F7	Patchbay
F8	CD-RW/	CD/Mastering
Shift + F8	SHIFT+CD-RW	Mastering Room
F9*	PROJECT	Project List
F10*	TRACK	Region or Phrase
F11*	EFFECT	Effects
F12*	UTILITY	Numerics
Pos1 + Shift	SHIFT+HOME/DISPLAY	LCD-Display/ V.Tracks
Arrow keys + Shift	SHIFT + Cursor buttons	Playlist zoom
Arrow keys	Cursor buttons	Moving the cursor
Space	PLAY/STOP	Start/Stop
Strg + S	SHIFT + S	Save Project
Ctrl+ Alt + Del	SHIFT + STOP	Switch off the VS
Tab	PAGE	Menu pages
Enter	YES	Execute function
Esc	EXIT	Abort function

* Only VS-2480

Before operation, the keyboard must be activated in the system menu:

Utility → System3 → *PS/2 Keyboard*

The "Keyboard Type" field additionally allows the localized settings of the Keyboard to be called up (e.g. PS/2 Germany).

Knob/Fader Assign

The VS-24xx models allow the →Faders* to be used for more than just setting levels. This concerns the →AUX Sends, but also a freely chosen mix parameter. Activate the function by using the indicated buttons:

VS-2400:	FADER ASSIGN
VS-2480:	KNOB/FADER ASSIGN

* VS-2480: Fader and PAN/AUX knob

In this state, the faders no longer act as volume controllers, but rather level the chosen AUX Sends. This is indicated by a flashing button. The numbered buttons 1~8 now specify the fader assignment to AUX Sends 1~8, while the 9/USER button can be used to select the definable parameter.

Selecting and Setting AUX Sends

The numeric buttons 1~8 are used to assign the required AUX Send bus to the faders. By default, the faders are assigned to AUX Send bus 1. Use "2" to set the levels for AUX bus 2, etc. To check this, CH EDIT or the display of several mixer channels can be called up (→MltChV, →TR F/P or →IN F/P).

Assigning a Mixer Parameter for the Faders

Press the 9/USER button to be able to set the selected mixer parameter using the faders. Start by selecting a parameter:

Utility → V.FDR/Usr → *FDR ASSIGN to** → F3 Param3 → "User Fader Assign"

In this field, select one mixer parameter for the input of values via the faders.

Application:

- Filter control (direct access to the "High Gain" parameter of all channels)

LA-2A

- Setting the noise gate (expander) Threshold;
- Assigning fader groups, etc.

VS-2480: "Knob/Fader Assign to"

Note
Fader editing may lead to a lot of confusion. It is therefore recommended to return to the level mode as soon as the AUX Send settings are satisfactory.

VS-2480:
On the VS-2480, →Knob/Fader Assign allows alternating between fader and encoder (Pan knob) entries for parameters:

System → Utility → Global Knob/FDR Assign Sw

Even though the faders may be an easier solution for setting certain parameter values, it is recommended to use the encoders as much as possible. In this way, the faders keep their level setting function and there is less confusion.

L

LA-2A
A →Plug-in from "Universal Audio". Thanks to its program-dependent operation (opto compressor), the LA2A, originally conceived for broadcast applications, was quickly adopted by professional recording studios around the world. It is mainly used for processing vocal, bass guitars, guitars and horns. Even at extreme settings, its highest frequencies remain sparkling, and its definition impeccable.

Fig.: LA2A vintage compressor (Universal Audio)

LA2A
Effect patch:	Plug-in
Connection:	Insert
Application:	Vocals, optional

The level meter's "Gain Reduction" setting allows the intensity of the compression to be controlled. This is determined by the "Gain" (onset of the level reduction) and "Peak Reduction" (compression ratio) controls. The "Limit" parameter enables the compression to be intensified. Due to its slow attack time (typical for the opto compressor principle) a hard limiting should not be expected.

Lack of Events!
This →Error Message means that the number of remaining →Events is insufficient to perform the requested operation.

Work-around:

- Save the Project (to discard the Redo data, →Project Store)
- →Project Optimize
- Delete Automix data

LANC
The "Local Application Control Bus System" is a standard for the transmission of →Timecode and machine control operations of video devices. It usually uses an "L" (sub mini jack, mainly Sony) or a 5-pin edit connector (Mini-DIN, usually Panasonic). A suitable →Synchronizer (e.g. →SI-80SP) can convert LANC signals to →MTC, which allows remote control of the VS from a DV camcorder or DV video recorder (→Video Recorder Synchronization). The battery-powered 4-channel recorder Edirol →R-4 (Pro) is fitted with a LANC converter.

LCD Display
In the same way as a VGA display, certain functions can be operated by using the mouse. Even the well known menu system used by the VGA screen is available:

Level Setting

Click → HOME

LED
A "Light Emitting Diode" is used to signal several operation states (most newer models can light in different colors).

Level
The level represents a measurement unit for the electric voltage and the related volume. It is indicated as an absolute level in Volt, or →dBu/dBV, i.e. the relative level in dB. Whereas the →Line Level is comprised of voltage levels of 100mV to 1V, →Microphone Levels have an extremely low voltage of ±2mV for dynamic microphones and up to ±20mV for condenser microphones. The levels indicated by the VS's →Level Meters refer to the digital maximum value of 0dB FS (full scale). The levels indicated by the VS's →Level Meters refer to the digital maximum value of 0dB FS (full scale). The VS allows levels (→Level Setting) to be increased or decreased using →Faders, →Attenuators, →Scenes →Dynamics, →Automix, →Normalize, →Phrase Parameter, etc.

Level Meter
LMs measure the levels of audio signals and are available for most operations on the VS. They also indicate whether or not a track contains data. Furthermore, they can be used to check routings. Except for the input and →Master Section signals, all LMs can be routed →Pre/Post.

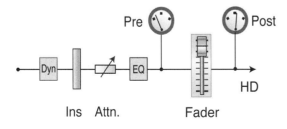

Fig.: Level metering of one channel

Note:
Pay attention to the influence of the →Dynamics processors, →Insert effects, →Attenuators and →Equalizers on the signal level. Even when set to "Pre", the level meters only indicate the input levels for the corresponding channel strips when their equalizers are flat. See the illustration below:

The mix bus can be monitored using the Master level meters. Even the monitor bus has its own meters.

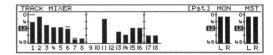

Fig.: Track level meters in the LCD (VS-2000)

The desired meters can be activated on the VGA screen by clicking a check box or using the function buttons on PAGE 1/3. See the table below for the corresponding names and F button assignments:

Section	VGA	F button (1/3)
Input section	INPUT	F1 (INPUT)
Input Mixer	IN1-xx	F2 (IN Mix)
Track Mixer	TR1-xx	F3 (Tr Mix)
AUX Master	AUX/DIR	F4 (AUX/DIR)
Direct Path	AUX/DIR	F4 (AUX/DIR)
Outputs	OUTPUT	F5 (OUTPUT)
FX Return	FX1-x RTN	F5* (FX RTN)

* PAGE 2/3

Signal peaks can be held in all level meters (→Peak Hold Sw).

Level Setting
"Level Setting" refers to the act of changing a signal's level. The VS provides several functions for this: →Sensitivity Knobs, →Faders, →Attenuators, →AUX Sends, →Phrase Parameters, etc. During →Recording, all input signals need to be set using the Sensitivity knobs (leave the associated faders at "0dB"). In the

case of the faders, positive values represent level boosts, whereas negative values represent level cuts. "0dB" indicates that the level is left as is (→dB).

Level Setting While Recording

L. controls the electric level to ensure a perfect balance at the time of →Recording and during →Mixdown by referring to the →Level Meters. Thanks to its 24-bit →A/D Converters and the internal 56-bit processing, the VS even provides an acceptable resolution for low-level signals. The average level of all signals should be about –12dB, with occasional signals peaks at –4dB. This leaves sufficient →Headroom for unexpected level jumps.

> Take the necessary steps to ensure that signal peaks don't exceed –4dB.

To avoid →Clipping, the 0dB mark should never be exceeded. This explains why the level meters use extremely fast response times of 1ms.

It is of utmost importance that the point where the level is measured is before or after the fader (→Pre/Post).

Light Pipe

The optical transmission of →S/PDIF- or →ADAT-format audio data is sometimes called L.

Limiter

An L. is in fact a →Compressor whose →Ratio is set to 1 : ∞ (or at least 1:100) and has extremely fast regulating times. This no longer reduces peak levels, but restricts them. All signal levels in excess of the Threshold setting are attenuated to the Threshold level. In the illustration, signal levels up to –10dB are left unchanged, whereas louder passages are attenuated. This creates a new headroom of ±10dB, by which the overall signal level can be boosted. The VS allows the channel compressors to be configured as substitute limiters by setting Ratio to "∞" and Attack to "0". The Release value should also be short (although level pumping should be avoided).

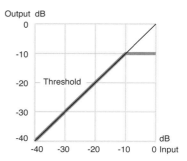

Fig.: Limiter threshold

One of the →Effect Patch, the →Mastering Tool Kit, provides a multi-band compressor followed by a limiter. The →DualComp/Lim can also be used as limiter. Still another possibility would be to use the Peak Limiter →Plug-in called →Urei 1176 Compressor (by Universal Audio).

Line Level

With voltage values between 100mV and 1V, Line level (unlike →Microphone Level) is considered a high level. This is the level of synthesizers, sound modules, CD players, etc. It usually requires little or no amplification. While →Unbalanced Line signals have a level of –10dBv (±0.3V), →Balanced connections use +4dBu (1.23V, like the VS's outputs), or even +6dBu (1.55V, German broadcast standard). This may require the use of special level converters.
Absolute voltage levels:
0dBu = 0.775VV
0dBv = 1.0V.

Linear/Exponential

→Gradation (p. 86)
→Phrase Parameter (p. 154)

Link

The VS allows adjacent channels to be paired using the →Fader Link (FDRLINK, only levels) and →Channel Link modes. The channel link mode includes all mixer parameters, i.e. the level setting, dynamics, effect sends/inserts, the attenuator, group assignments, automix, etc. – even the →Track Status and →V-Track number.

List
→Project List (p. 167)

Listening Level
The listening level can be set using the MONITOR knob or on the power amplifier connected to the →Monitor outputs. Do not use the →Master Fader. As a rule, low listening levels help establish the correct level balance. Higher levels should only be used to briefly check the bass content of the mix.

Live Mix
An L. involving the VS requires the appropriate number of inputs and the required outputs towards the PA and monitoring systems. Use the →Input Mixer section for plain live mixing. The →Track Mixer channels are only needed to record the live signals.

Inputs
All VS models provide eight →Analog Inputs for live applications (they are configured as →XLR sockets that need to be connected to the multicore cable). The VS-2480 has eight additional balanced jack inputs for which appropriate jack-to-XLR adapter cables are needed. By adding an →ADA-7000, the number of microphone inputs is increased by eight.

Tip
Consider using an external device's digital outputs whenever possible, as this frees up the equivalent number of analog inputs on the VS-24xx (→Recording of digital Signals).

Outputs
All of the VS's outputs are configured as stereo jacks. Because the power amplifier connected to them usually has XLR inputs, remember to use →Balanced jack-to-XLR (male) adapter* cables. Use the MASTER outputs to feed the PA system. The AUX outputs can be used for the →Monitor Mix.

> * VS-2000: AUX outputs configured as phono/RCA sockets

Recording Individual Signals
Simultaneously recording input signals while taking care of the FOH mix is relatively easy. Set the INPUT MIXER faders to 0dB and balance the signals with the →Sensitivity Knobs. Level, filter, dynamics* and Send settings for the monitor system must be performed on the →Track Mixer level to avoid affecting the signals being recorded.

> * AUX signals are sourced after the →Dynamics processors. The processors, therefore, have a great influence on the monitor signals. Try to avoid using the compressors, because they are likely to cause feedback problems.

Recording The Mix Bus Signal
The following routings are available for recording the stereo mix bus signal:

1. Analog Patching
This approach can be easily implemented without changing the internal routings. The disadvantages, however, are that it requires additional analog inputs and the associated →D/A Conversions and →A/D Conversions.

- Use two phone cables to connect the monitor output to two inputs.
- Route those inputs to two tracks and activate them for recording.
- The Monitor knob and the channels' gain knobs set the recording level.
- Any additional changes to the signals can be performed on the INPUT MIXER level.

2. Bouncing
One factory preset for internal bouncing also routes the inputs and →Effect Returns to a stereo track.

- On the →EZ Routing 1 page, press F5 (Load) and select the "Bouncing" preset.

Live Recording

- Any additional changes to the signals can be performed on the INPUT MIXER level.
- Channel 23/24 on the TRACK MIXER level can be used to set the overall level.

Tip
Projects whose →Record Mode is set to →CDR, record image files that can be burned to a CD right away.

Master Effects
If an entire effect board is still available, the effect →Mastering Tool Kit can be advantageously inserted into the Mix bus. Doing so allows certain frequency deficiencies to be corrected, or the mix signal to be compressed.

Live Recording
As a compact device, the VS is ideally suited for L. Except for the →Monitor Mix, which is unnecessary here, →Recording is the same as by normal studio jobs. As a VGA screen normally won't be connected, all operations need to be performed via the internal LCD. This may require some practice, as one familiar interface is missing. Connect the FOH mixer's "Direct Outs" to the VS (or use splitters, which are usually rather expensive). In some cases, using the VS for the FOH mix may be more practical (→Live Mix). Set the levels of the input signals (→Level Setting While Recording) and check the settings with a pair of headphones. At the time of recording, it should be possible to merely monitor the levels of the signals being recorded. Monitoring settings should be performed via the →Track Mixer level, because its settings do not affect the signals being recorded.

Lo-Fi Processor
This →Effect patch deliberately reduces the quality of the audio signals being processed by reducing their →Sample Rate and →Word Length. A resonant filter that follows it allows classic synthesizer sweeps, while the →Noise Suppressor blocks unwanted noise during pauses.

Bit Crusher
In addition to lowering the →Sample Rate, which leads to a reduced frequency range, lowering the →Bit Rate usually produces noticeable digital distortion.

Realtime Modify Filter:
The parametric resonant filter can be used for experimental sounds, as a synthesizer filter, or as a conventional HPF, LPF or BPF.

Lo-Fi Processor	
Effect patch:	P027
Connection:	Insert
Application:	Experimental

Tip
Only the VS-2480 provides an additional →Filter. But if all other "Lo-Fi Processor" effect blocks are deactivated, this effect can be used as a →High-Pass filter on the other models.

Load
1. Sometimes erroneously used to signify the preparation of a →Project (→Project Select (p. 169)). This only loads the Project settings. The audio data resides on the hard disk and is played back from there.
2. Refers to recalling one of the VS-24xx's routing templates (→EZ Routing).
3. Refers to recalling one of the →Effect Patch or →Plug-ins for an effect processor.

Locator
To immediately jump to a chosen time position with the touch of a button, the VS provides 100 Locator storage places. These are divided over 10 banks of 10 locators and use absolute assignments of Locator numbers and stored time positions (unlike →Markers). The VS-24xx models have a numeric key pad that can be used to access the 10 memories of the currently selected bank. The VS-2000, on the other hand, uses the STATUS buttons for this.

Note:
Locator positions are not shown in the →Playlist.

Locator/Scene Type

1. VS button

Locator positions can be stored and recalled using the VS's buttons. To store a position, move the →Position Line to the required location (→Positioning).

- The LOCATOR function must be active (default).
- Press a 0~9 button that doesn't light to assign the →Current Time to it.
- Pressing a lit 0~9 button jumps to its stored position.
- To clear a Locator's stored position, hold down CLEAR while pressing its button.

Locator Bank: SHIFT + LOCATOR
The lit button indicates the bank that is currently selected. Pressing a flashing button activates the corresponding Locator bank.

VS-2000: First press LOCATOR. The "STATUS" buttons now act as Locator buttons. Locator bank: Hold down LOCATOR for at least 2 seconds, then use the dial or the mouse to select the bank + ENTER.

Save Mode

The Locator's Save mode, also available in the LCD, offers a clearly arranged method to confirm a positional change before it is actually performed.

Utility → Global*1 → "Locator/Scene Type"

Start by activating the Locator mode (the LOCATOR button lights). The VS now asks for a two-digit Locator number to be entered. Confirm the entry by pressing ENTER to jump to that position, after which the VS leaves the Locator mode.

* VS-2000: "System" rather than "Global". Confirm with [YES].

When an empty locator is selected, a message appears asking this operation to be confirmed ("Regist..?"), after which the current position is saved.

2. VGA screen

Click  in the right hand side of the →Measure Bar to open the Playlist button field. Via "Locator" the Locator field displaying the Locator positions of the currently active bank can be selected:

Fig.: Locator in the Playlist field

- Click one of the empty (dark) Locators to store the →Current Time to that memory.
- Click a previously stored (lit) number to move the →Position Line to that position.
- Holding down the VS's [CLEAR] while clicking a Locator No. erases that memory.

Locator bank: Click the bank row

3. Editing Locators

A special Locator Utility page displays a table where locator positions can be set, corrected and named.

Menu: Utility → Locator
VGA: Click [LOC].

Clicking a time or measure field in the table enables its value to be changed. F3 (GetNow) creates a new Locator at the current position and F2 (Clear) deletes the entire line. To jump to a given Locator position, select the corresponding entry and press F4 (Go To). →Name Entry becomes available F1 is pressed (Name).

Locator/Scene Type

As an unintentional →Scene change would overwrite any settings, the Scene saving recall operation "Save" instead of the default "Quick" can be chosen:

Utility → Global1 → "Locator/Scene Type"

Loop

With the setting "Save" a table provides a second safety mechanism. To make a selection, enter the two-digit Locator or Scene number, or click the corresponding row with the mouse. Press [YES] to confirm the choice, or [CLEAR] to abort the operation (VS-2000: "System" rather than "Global").

Loop

A "loop" is a passage which is played back repeatedly. This can be used to practise difficult riffs or solos. It is also possible to loop an entire song while →Mixing or mastering it. The FROM position defines the beginning of the loop. TO represents the end of the passage. Loop points can be set at anytime (during playback/recording and after stopping it)

> VGA: Right-click the Measure Bar → Loop
> VS: Hold down LOOP + FROM or TO

Note
Designed to assist while working, the Loop function cannot be used for rhythmic purposes such as drum loops because there will always be a short pause between the end and the beginning of a loop.

Activating The Loop
To activate, stop the Project and press the LOOP button, or click the Loop icon on the VGA screen. When playback starts, the →Timeline jumps to the FROM position and looped playback starts.

Editing The Loop Points
- Setting new loop points automatically overwrites the previous ones.
  To delete a loop point, hold down [CLEAR] and click it on the VGA screen.
- Clicking and dragging a Loop icon with the mouse in the VGA
- Loop positions can be entered numerical via the "Auto Punch/Loop" menu. These functions can also be used to register the current position, change, or delete a setting.

> VS: SHIFT + LOOP
> VGA: SHIFT + click 🔁

→Loop and Auto Punch (p. 108) can be used to rehearse a passage before committing it to hard disk.

Loop and Auto Punch

→Loop and →A.PUNCH can be used in combination. Repeatedly playing back a difficult passage allows parts to be practiced as often as necessary. When ready, press [REC] (or a →Foot Switch) to start (and stop) recording between the Punch IN/OUT locations, while loop playback keeps running.

Loop Effect
→Send/Return Effect (p. 204)

Loop Effect Assign
The →AUX Busses can be assigned to the internal effect processors (not on the VS-2000) using the processor's →FX Return Channel, the VGA's →TR F/P page or in →EZ Routing. The source can also be a →Direct Path. In most instances, the preset assignments will not need to be changed:

AUX bus 1 → Effect 1
AUX bus 2 → Effect 2, etc.

Application:
Assigning different send busses to an effect using other AUX busses.

Tip
When installing a →Plug-in →Effect Board, already available effect expansion boards need to be moved to different slots (not on the VS-2000). Effect assignments to AUX busses 1 and 2 now refer to the plug-in effect board and therefore no longer work as expected. Therefore "Assign" allows effects 3 and 4 to be assigned to busses 1 and 2 (useful for previously recorded Projects).

Loudness

A signal's loudness is usually not something that can be inferred from watching its meter indication. Whereas digital audio systems have to show even extremely brief peaks, the human ear reacts more slowly, and has a tendency to consider mid-frequency sound pressure levels louder than others. Volume levels measured using technical devices should never be confused with subjective loudness impressions. This is especially important for →Mastering operations that usually involve using a →Compressor and →Limiter to squeeze the dynamic range.

Low

The →Equalizers of the VS's mixer channels provide a L. shelving band allowing to boost and cut the bass range.

Low Mid

On a VS, the lower peaking filter of an →Equalizer section is called L.

Low-Pass

The VS-2480's →Equalizer provides an additional multi-purpose filter that can be set to "LPF". It only attenuates frequencies above its cutoff frequency (at 12dB/octave), while the range below it passes through unaffected. (→High- and Low-pass filters of the VS-2480).

Tip
The filter can also be used for synthesizer-like sweeps by emphasizing its cutoff frequency using the "Q" parameter.

M

M16

The M16 →Record Mode represents the uncompressed 16-bit format.

M24

The M24 →Record Mode refers to the linear, uncompressed 24-bit format. Note that the number of recording and playback tracks decreases when this mode is selected. In most instances, the more economic →MTP will suffice.

Mark

Certain →Project List operations require a confirmation with "Mark". This is also necessary for track assignments while →Editing. Desired tracks need to be marked in the check box.

Marker

Similar to →Locator positions, markers are designed to quickly jump to given →Timeline positions. Given the impressive number of markers, they can be abundantly used. They can also be named and therefore used to help structure a Project.

Application

- Structuring and locating song parts already at the time of recording.
- Using them for navigation within a project and while →Editing.
- Belated tempo indication for songs, which were recorded without click: →Sync Track Convert
- Changing the project's arrangement: →Region Arrange
- Use as →CD Markers for →CD Burning.

1. Setting Markers
Move the timeline to the desired position, then proceed as follows:

Mouse:	Right-click the Measure Bar ➔ Marker
VS:	TAP button
VGA:	Click TAP

2. Recalling A Marker
When selecting a Marker, the timeline jumps to the memorized location.
Markers can be selected in several ways:

2.1. Previous/Next Buttons
These buttons recall the previous or next Marker and move the timeline to that position*.

MARKER Button

VGA: SHIFT + ⏮ ⏭
VS: SHIFT + PREVIOUS or NEXT

Or corresponding →Phrase edges, see →Previous/Next Switch.

2.2. In The Time Display

The →Time Display's Marker field allows Markers to be selected by dragging the mouse or turning the dial. The →Current Time and the position line move to the Marker's position.

Fig.: Marker field in the time display

2.3. 'Marker Utility' Page

This page displays a table that contains all Markers to be selected and edited (see "4. Editing Markers").

2.4. Marker Display

This screen displays all markers in a table for Marker selection and editing:

VS: SHIFT + F6 (Marker)
VS-2480: MARKER button
VGA: Click MRK

Here Markers need to be selected by entering their three-digit number.

YES: Jumps to the marker position
CLEAR: Deletes the Marker

3. Deleting

Markers can be deleted via VS, graphically and in the Utility's Marker screen.

VS: Locates to the Marker position, CLEAR +TAP
VGA: CLEAR + click TAP

Deleting all Markers of a Project:

Mouse: Right-click the Measure Bar → Clr All M.
VS: SHIFT + CLEAR + TAP

4. Editing Markers

Marker positions can be corrected by moving the Marker icon on the VGA screen. A special Marker Utility page displays a table where Marker positions can be set, changed and named (→Name Entry).

Menu: Utility → Marker
VGA: SHIFT + click MRK

Clicking a time or measure field in the table allows its value to be changed. F3 (GetNow) creates a new Marker at the current position. F2 (Clear) deletes the entire line. To jump to a given Marker position, select the corresponding entry and press F4 (Go To). →Name Entry becomes available when F1 (Name) is pressed.

5. CD Markers

CD markers are used to identify the different CD tracks that can be selected on regular CD players. These markers are to be set independently of the VS's regular Markers (→2..CD Markers (p. 36)).

MARKER Button

Only the VS-2480 provides a dedicated button for recalling the Marker display (see "2.4. Marker Display").

Marker Stop

M. means that playback will stop at the next Marker position.

Application

- For live performances, this function allows playback to be stopped automatically at the end of each song.

Utility → Play/Rec1 → Marker Stop

Massenburg EQ

The "High Resolution EQ" →Plug-in by Massenburg* Designworks is a high-quality audio tool (→Parametric Equalizer). It only runs after installing a →VS8F-3 →Effect Board, it uses the processing power of both its processors.

The →Sample Rate of the signal to be processed is upsampled to 88.2 or 96kHz (even if the Project uses a lower frequency) to ensure perfect phase stability and a near-analog sound, which is especially important for high frequencies. Settings are possible within a range of 10Hz and 32kHz.

* *George Massenburg presented the first parametric equalizer at the 1972 AES convention.*

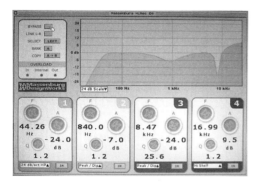

Fig.: The Massenburg parametric EQ (plug-in)

Load the →Plug-in into an odd-numbered effect processor and use "Patch" to select an EQ preset. Press F4 (Edit) to jump to the edit page. This stereo or dual-mono EQ provides 4 bands per channel with overlapping center frequency ranges whose levels can be boosted/cut by 24dB. All bands can be used as →Peaking or →Shelving filters. The slopes of the highest and lowest bands are adjustable. There are two banks with a possible copy function. Though the display has a somewhat slow response time, the audio data is processed without flaws.

Massenburg HiRes EQ

Effect patch:	Plug-in
Connection:	Insert
Application:	High-end

Tip
This effect enables the independent use of both Mono-EQs blocks in two mixer channels. In a serial connection, however, an 8-band Mono-EQ is available.

Master Clock

Master
When two devices are synchronized, one of them acts as master and sends →Timecode signals to the other (the →Slave), which in turn is operated by these time references.

Master Bus
→Mix Bus (p. 128)

Master Clock
In a digital audio system, the →Clock generator provides the →Sample Rate, which is necessary to synchronize two ore more digital systems. While this is usually generated internally by the VS, →Recording of digital Signals is only possible if the clock is synchronized with the external device. The menu for this can be selected as follows:

VS:	Utility → Project Parameter → Master Clock
VS-2000:	Utility → Digital → Master Clock

On the page "Master Clock", the clock source, Dither function and copy-protection flag can be selected.

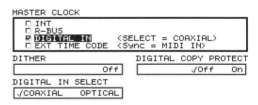

Fig.: Clock settings (VS-2400)

Select the digital input through which the digital signal will be received (R-Bus, coaxial, optical, or DIGITAL IN on the VS-2000). When the clock signal is recognized, the following message appears:

"Digital In Lock"

If the external device is also used as →MTC master for →Synchronization, select "EXT TIME CODE". In this mode, the VS derives the clock information from the incoming time code and uses it for digital and tempo synchronization. Only the VS-2480 provides a separate →Word Clock input.

MASTER EDIT
Select the M. menu to edit the parameters in the VS's →Master Section:

VGA:	Mixer → Master Edit
VS-2400:	MASTER EDIT
VS:	SHIFT + MASTER EDIT

The level and Pan of the Monitor and →Mix Busses as well as the →Aux Master/→Direct Paths can be set here. Additionally, this menu allows effects to be inserted into the MASTER bus and the →Output Assign parameters for the busses to be set.

Master Effect
Mix Bus signals can be processed by →Insert effects connected to the →Master Section. The insertion point is located before the bus fader. Doing so allows the entire mix rather than individual channels to be processed. For reasons of →Effect Economy, Mix bus processing (→Mastering) is usually only applied in a separate stage.

VGA:	Mixer → Master Edit → F2 (FX Ins)
VS:	SHIFT + MASTER
VS-2400:	MASTER EDIT
VS-2000:	SHIFT + F3 (Effect)

Inserting effects into the MASTER section follows the same pattern as working with the →Mastering Room.
The effects normally used for MASTER processing are:
→Mastering Tool Kit, →Massenburg EQ, →T-Racks Mastering, →Speaker Modeling, →Reverb (only stereo, to add spaciousness, be sure to choose a proper MIX setting).

Master Fader
The M., located at the end of the →Mix Bus (→Master Section), is used to set the overall Mix level. In the →Mastering Room its setting has a direct influence on the data of the →Master Tracks. It should therefore never be used to set the →Listening Level. The "MST" →Level Meters indicate the Master bus level. The M. can be automated like any other channel via →Automix).

MASTER OUT
The M. sockets transmit the →Mix Bus signals (fixed assignment*) sourced after the MASTER fader.

Not →Track Direct Out, not on the VS-2000

Master Section
The M. is used to set the level and stereo balance of the →Mix Bus, the Monitor Bus and the →Aux Master/→Direct Paths. Master effects are able to be inserted and the →Output Assignment of the bus signals can be routed.

VGA:	Mixer → Master Edit
VS-2400	MASTER EDIT
VS-2480/200	SHIFT + MASTER EDIT

F1 View
Allows the →Panorama of the Master and Monitor busses as well as the →Pre/Post and →Link settings to be set for the Bus Masters of the →AUX Busses and →Direct Paths.

F2 Insert
→Inserts are located before the MASTER fader and Pan control. The effects are identical to those of the →Mastering Room.

F3 Output Assignments
This page is the counterpart in table form, to the graphical Output Assign page of →EZ Routing. Changes made here are reflected on the EZ Routing page as well.

Level Metering
The MASTER section's →Level Meters are sourced post-faders and therefore indicate the level of the signals transmitted to the →MASTER OUT sockets (or the recording level of the →Master Tracks).

Mastering

M. refers to the operations performed to achieve the desired →Loudness and overall sound of the Stereo Mix signals. This is usually the last stage before burning a CD, and thus the last chance to perfect the overall sound of a mix. Next to mixing, the mastering stage is one of the most demanding processes in music production and requires some experience and knowledge. Possible shortcomings of the mix cannot be fixed here, rather the overall sound can merely be enhanced by making it tighter or boosting/attenuating frequency ranges that appear too soft/too prominent. It is normally impossible to correct the channel balance (but see also →4. Correcting A Mix Without Redoing It (p. 127)). The following Mastering Tools are provided: a →Multiband Compressor, a limiter, an →Equalizer, and effects that allow the stereo image to be broadened (→T-Racks Mastering). Although in the VS mastering directly during the mix would be possible, (→Master Effect), this is usually not done for reasons of →Effect Economy, Changed settings, listening conditions and, in particular, the Mastering operation, demand full attention and 'fresh ears'.

Mastering With The VS

The stereo mix recorded in the →Mastering Room is the material being worked on. Here, they are called the →Master Tracks (this could become confusing, therefore it is better to call them "Mix Tracks"). Because these tracks for the following mastering operation were not (!) recorded in →CDR mode, they are able to be played back like normal tracks. The tool usually used for mastering is the effect →Mastering Tool Kit. Simple operations can also be carried out by using the channel's →Dynamics and →Equalizer.

1. Activating Tracks

The last two →Track Mixer channels (VS-24xx: 23/24, VS-2000: 17/18) need to be linked with →Channel Link. By selecting →V-Track 16 (or the V-track pair selected for recording the mix), they become audible. The remaining tracks are no longer needed and should therefore be muted (or deactivated via →Track Status).

2. Inserting The "MTK" Effect

Assign "Mastering Tool Kit" to effect processor 1 (or any other odd-numbered processor) on the "Effect" page (→Effect Patch)*. The "P232 [PreMastr]" patch usually provides a good stating point. →Insert that processor into tracks 23/24. This should have an immediate effect on the Mix Tracks.

> * Alternatively, the "MTK" →Plug-in can be selected, which requires the power of only one processor.

3. Reference Title

Prepare recordings by other artists whose style and/or sound is similar to the desired effect to be achieved. This sound can act as a guide. Frequent →A/B Comparisons should pinpoint the aspects which still need work (frequency response, etc.). Doing so can compensate, middle rate →Studio Monitors, unfavorable listening conditions, lack of experience, listening weariness etc. The first step is usually to adjust the frequency response (high and low frequencies as well as the critical mid range) of the material. By repeatedly toggling between the Reference CD and your own Mix and subsequently modifying the Mix sound, a gradual approach to the reference sound becomes possible. For an easy A/B comparison, it is necessary to provide an ergonomic change-over switch. The reference material can be used by recording it onto VS tracks or by playing it back live.

3.1. Recording The Reference Material

CD passages, which are to be used as a guide can be recorded to tracks 21/22 (next to the Mix Tracks). If possible, record the material in the digital domain (at 0 →dB), because this will help to judge the →Loudness of the mix when using the same levels. To compare, alternate between the reference material and the mix. The Diagram shows the inserted MTK patch in the mixer channels of the Mix Tracks and the →Speaker Modeling (SPM) as an additional →Master Effect.

Mastering

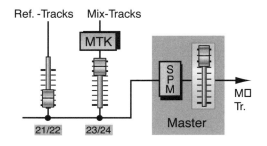

Fig.: Mastering setup on the VS

"Speaker Modeling" emulates the frequency response of various monitor speakers. This allows a comparison of the mix to the reference material through different sounding simulated speaker boxes. Although this algorithm was intended for specific monitors, it can nevertheless be used as another listening variation. Especially the simulation of small speakers* should allow an analysis of the mid range (the most important frequency range). Use →Bypass to switch the speaker modeling algorithm on and off when necessary.

* *Small Cube, Small Radio, Small TV*

3.2. Live Playback Of A Reference CD

A commercially available CD can also be used as a reference source. Connect the CD player to the →Input Mixer and set the correct level.

Tip
The CD player's level can be set more easily if a fader →Group is produced via faders 21/22 on the TRACK MIXER level. This has the advantage that the faders of the reference source and the mix being compared lie next to each other.

4. Filtering The Mix Tracks

While the exact frequency contents of the masters typically depend on the material recorded there are some general rules for certain frequency ranges. At the mastering stage, the EQ is often used for sweetening purposes, i.e. to boost or cut relatively broad frequency ranges. This allows a Project's sound to liken that of tracks mastered at an earlier stage, or to correct the bass response etc.

Frequency	Effect
< 80Hz	Foundation, warmth
≈250Hz	Attenuating rumbling frequencies
≈2kHz	Make the sound less nasal
≈4kHz	Lower for a velvety high range
> 10 KHz	Boost to add some "air"

Tip
Judging the bass contents is relatively difficult in small listening rooms. Consider listening from the hallway or an adjacent room from time to time.

Knowing the key of the song also helps to pay special attention to the relevant frequency bases. See the table below.

	ctr.	gr.	sm.	1	2	3	4
C	32.7	65,4	131	262	523	1046	2093
C#	35	69	139	277	554	1109	2217
D	36.7	73,4	147	294	587	1175	2349
D#	39	78	156	311	662	1245	2489
E	41.2	82,4	165	323	659	1318	2637
F	43.6	87,3	175	349	698	1396	2794
F#	46	93	185	370	740	1480	2960
G	49	98	196	392	784	1568	3126
G#	52	104	208	415	831	1661	3322
A	55	110	220	440	880	1760	3520
A#	58	117	233	466	932	1865	3729
H	61.7	124	247	494	988	1975	3951

Fig.: Note frequency table

5. MTK And Alternatives

The "MTK" effect provides all tools needed for mastering applications as separate blocks. The "PreMastr" preset is usually a good place to start. Depending on the material to be mastered, patches 231~ 241 can also be used for Mix Track processing. See →Mastering Tool Kit (p. 117) for a discussion of the available parameters. Further mastering effects are available as →Plug-ins: →Massenburg EQ (p. 110), →T-Racks Mastering (p. 220): Equalizer and dynamics. "Dynamics" is the only effect that can broaden the "St. Image", which is often used in the mastering stage.

6. Recording The Mix To Be Mastered

Recording the processed Mix Tracks is performed in the →Mastering Room. The resulting tracks are called →Master Tracks. Select an empty V-Track pair and at this time activate the →CDR mode in order to subsequently burn the material faster onto a CD. See →3. Recording the Mix (p. 126) for details.

Mastering Room

The "Mastering Room" displays all playback and recording settings of the →Master Tracks on one single menu page. In addition to the playback tracks, the master tracks also record the input and FX Return signals. Additionally, an effect can be →Inserted into the Mix Bus, a special →CD Burn mode can be activated, markers automatically set and the Master Tracks positioned:

VGA: CD-RW/Mastering → Mastering Room
VS: SHIFT +CD-RW/MASTERING

Fig.: 'Mastering Room' (display page)

The "Mastering Room" menu appears above the Playlist, allowing the master tracks (after scrolling down) to be checked and even edited:

Recording The Master Tracks

1. Activate the "Mastering Room".

   VGA: Click "Mastering Room" or [MSTR]
   VS: Use the dial

   The CDR-W/MASTERING button flashes.

2. Select the "Record" status.
   Click ● or the STATUS button PLAY. Only after activating the "Record" status can the Project tracks be played back. In "Play" mode, only the Master Tracks (see below) will be heard.

3. Select the V-Track pair to be used as →Master Tracks (stereo). There are 16 V-Tracks available and therefore, 16 different M. versions could be created.

4. Specify the master track's position (see the table below)

5. If necessary, generate automatic →Markers at the beginning of each CD track ("Auto Marker").

6. Insert the →Master Effect, if necessary.

7. Activate →CDR to burn a CD. This saves creating an →Image File, which makes the process faster. Such CDR master tracks can only be played back in the Mastering Room and never together with regular tracks (unless the entire Project uses CDR mode).

Application	After Record	CDR
Track at 0:0:0	to ZERO	free
Subsequent CD tracks	to last Phrs: 0 s / 2 s / 4 s	On
Intermediate mix	stay HERE	Off
Mix correction without new mix*	stay HERE	Off
Effect economy	stay HERE	Off

* →4. Correcting A Mix Without Redoing It

The following illustration shows the signal flow for recording Master Tracks. To record the Master Tracks, locate the desired position,

Master Insert with External Effect

then press REC + PLAY to start recording. All signals sent to the →Mix Bus (tracks, input signals, FX Returns) are recorded to the M. Tracks.

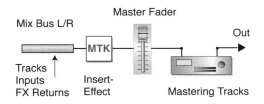

Fig.: Master tracks at the time of recording

The MASTER fader sets the bus level (and fades).

Playing Back The Master Tracks

Activate the Mastering Room, then proceed as follows to play back the master tracks:

1. Select the Mastering Room's PLAY status.

   VGA: Click the red ● icon or REC to switch to "Play" status.

   VS: Use the dial

2. Jump to the beginning and start playback.

The Mastering Room's PLAY status deactivates the mix bus. Only the MASTER fader can be used to set the master track's level. There is no insert effect.

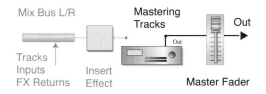

Fig.: Master tracks during playback

The Mastering Room's "Play" mode only allows the master tracks to be played back. All other audio signals are muted. There is therefore no need to set the TRACK and INPUT MIXER faders to –∞.

Master Insert with External Effect

Though there are no insert sockets that would allow an external effect processor to be inserted, special routing systems can be used to establish such a connection on an analog (sometimes even a digital) level. This allows high-end valve compressors, equalizers, exciters and other psycho-acoustic processors to be used while →Mastering. If taking advantage of this system, be aware that it will block two tracks (unlike the →Master Tracks). These tracks are used to record the processed bus, and are therefore no longer available for regular tracks.

Analog

Thanks to the 24-bit converters' high quality, using external analog devices usually doesn't lead to signal degradation. Use the following connections:

VS MASTER OUT → effect processor's inputs
Processor's outputs → VS inputs 1/2 (for example)

Activate the VS's →Link function for (e.g.) input channels 1/2, disconnect them from the mix bus (→MIX Switch) and set the faders to 0dB. Next, route input channels 1/2 to the desired recording tracks (23/24 for example, see "EZ Routing") and disconnect them also from the Mix. Engage recording for those tracks and link channels 23/24.

Digital

On the VS-2480, using external processors is fairly easy. Digital signals can be sent and received using the →Master Clock without any sample rate converter. On the other models, at least one signal flow direction can be handled in the digital domain. See →External Effect Processors for the wiring. Routing is the same as for analog mode.

Monitoring

For this application, the →Record Bus is routed to the →Monitor Bus. This allows the signals sent to the recording tracks to be monitored via MON (even headphones can be used). Leave the MASTER fader at "0dB".

Recording And Playback
→Recording to tracks 23/24 is the same as for any other application. The recording level is influenced by the external effect processor, the VS's MASTER fader, the Gain controls and channel faders 1/2 (for example).
For playback, the Monitor bus needs to be routed back to the Master bus. Listen to the newly recorded tracks in isolation (→Solo, etc.).

Mastering Tool Kit
The VS provides a special effect for →Mastering purposes called "MTK". Because of its numerous effect blocks, it can be used almost universally. A certain experience, however, is required for using the MTK.

Fig.: MTK (plug-in version, VGA)

Mastering Tool Kit	
Effect patch:	P231~249
Connection:	Insert
Application:	Mastering

The internal MTK requires the power of two processors and therefore needs to be assigned to an odd-numbered effect. The following figure shows the essential MTK blocks:

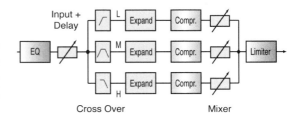

Fig.: Mastering Tool Kit (block diagram)

Equalizer:
The →Equalizer serves to cut/boost selected frequency bands. Though similar to regular channel EQs, it allows a →Shelving or →Peaking curve to be selected for the highest and lowest bands. Please note that the 3-band compressor dramatically influences the frequency range of the signal. See →Mastering (p. 113) for how to use the Mastering equalizer.

Bass Cut:
Frequencies below 20Hz are usually inaudible and only use up amplifier power. Use the →High-Pass filter to eliminate unnecessary low frequencies.

Enhancer:
The →Enhancer boosts the high frequency range (based on the *Sens* setting). Be careful not to overdo it.

Input:
InGain: Boosts the signal level to an appropriate value for the expander and compressor blocks. Dependent on previous EQ settings, be careful of the level as the signal may start clipping. In most instances, the level can be boosted quite a bit though.
D-Time: This induces a short delay, which allows the expander and compressor to react quickly enough to catch an incoming transient (Look Ahead). For mastering purposes, consider using the highest value. For live material, avoid exceeding 1ms.
Split Point: This cross-over divides the frequency spectrum into three bands that can be processed separately.

Master Tracks

Expander:
The →Expander is mainly used to suppress background noises. Its 3-band structure allows the restriction of its gating operation to the frequencies where the noise occurs (e.g. hiss induced by high compressor settings). The E. is not active in most presets.

Compressor:
The 3-band →Compressor is the most important part within the MTK. Unlike single-band compressors, it provides three compressors that process different frequency ranges for optimum compression across the spectrum. A high-level bass drum, for instance, usually only triggers the bass compressor, leaving the medium and high ranges unaffected. This division allows for stronger processing without unpleasant pumping.

Mixer:
After expanding and compressing the signals, the three frequency band levels can independently be set by boosting up to 6dB – or attenuating by up to 80dB. This function is similar to a 3-band EQ with fixed center frequencies. Extreme attenuation may yield interesting, experimental results.

Limiter:
The →Limiter keeps signal peaks from exceeding the Threshold level. It is not activated in some presets.

Out:
"*Soft Clip*" is an effective tool to soften digital distortions. When a digital signal exceeds the 0 dB, it's waveform changes to squares - the signal is clipping. Soft Clip rounds the square edges to create warm, saturated mastering effects. The MTK's "Softclip" parameter can be used to create additional headroom, allowing the overall volume to be increased. The level parameter at its output is usually only needed when the MTK is inserted into a mixer channel.

Sample Rates And MTK
The internal MTK is not available for Projects with a →Sample Rate in excess of 48kHz. The →Plug-in MTK, on the other hand, is not subject to this limitation. The MTK plug-in supplied with the VS8F-3 supports sampling rates up to 96kHz. For rates up to 48kHz, two MTKs can be used per processor.

Master Tracks
In the →Mastering Room, the →Mix is recorded to two empty →V-Tracks* of the last two tracks (called "Master Tracks"). Up to 16 different mixes can be recorded. The Master Tracks record the TRACK MIXER, INPUT MIXER and →Effect Return signals. When the Mastering Room is active, the M. appear below the "regular" tracks in the Playlist. Except for the VS-2000, It is necessary to scroll down to see them.

> *If the destination V-Tracks already contain material to be kept, copy it to other tracks before proceeding.*

Application:
- Recording the →Mix (p. 125)
- Image file for →CD Burn (p. 35)
- →Mastering (p. 113)
- Intermediary mixes, track →Bouncing (p. 30),
- setting track levels after the mix proper (!).

Tip:
- Master tracks not recorded in CDR mode can be used as regular V-Tracks at a later stage. Interim mixes of tracks and effects (→Bouncing) require no additional →Routing: just set the levels of the tracks to be bounced ("what you hear is what you get").
  After activating the Mastering Room, switch on →Channel Link for the last two channels (VS-24xx: 23/24) and select the →V-Track pair to be recorded to.
- Recording Live Signals To Stereo Tracks:
  Given that the M. also records signals coming from the inputs, use CDR mode to record a Master Track right away and then burn it onto a CD-R.

Measure Display

Editing In The Mastering Room
→Editing the Master Tracks directly in the Mastering Room is only possible in the VS-2000. For the other models it is carried out by editing the corresponding V-Tracks:

1. Activate "PLAY" in the Mastering Room.
2. Switch on →Channel Link for tracks 23/24.
3. Select →V-Track 16 (or the V-Tracks selected in the Mastering Room).
4. Edit the original V-Tracks. The CDR master tracks change accordingly.
5. Start playback to check the result.

Only M. that weren't recorded in CDR mode can be edited like regular tracks outside of the Mastering Room. Be sure to select the correct V-Tracks of the last two tracks.

McDSP ChromeToneAmp

Similarly to the VS's internal →Guitar Amp Modeling algorithm, this →Plug-in simulates the sound of an analog guitar amplifier.

ChromeToneAmp	
Effect patch:	Plug-in
Connection:	Insert
Application:	Guitar

Due to the high-impedance →Guitar Hi-Z input (which allows directly plugging in an electric guitar) and this amp algorithm, a real guitar amp can often be spared. ChromeToneAmp works for both stereo and two independent mono sources. The effect blocks can be used as follows: Low Cut (→High-Pass); →Gate; →Compressor/Sustainer; Distortion (see below); →Equalizer; →Reverb.

Fig.: ChromeToneAmp (McDSP)

Distortion
This block sets the basic tone. Select the algorithm (DIST1~DIST5) that yields the necessary degree of distortion. DIST BP simulates a fixed wah pedal that is used as band-pass filter. DRIVE/FREQ allows a selectable frequency to be boosted or cut. AMOUNT sets the amount of distortion. This is also indicated by a yellow LED.

Alternatives:
- →Guitar Amp Modeling (p. 87)
- →GuitarMulti (p. 88)

Measure

In the →Time Display, →Tempo Map, the →Grid function and the →Rhythm computer, M. refers to a subdivision in musical bars. This is only useful if the →Metronome is used while recording a new Project.

Measure Bar

The Measure Bar displays and allows the →Marker, →Edit Points, →Punch In/Out icons and →Loop to be set. Right-clicking the gray field above the →Playlist opens the pop-up menu with the available entries. The subdivision indicates the bars with respect to the Project tempo.

Fig.: Measure Bar

Measure Display

The →Time Display shows the current time both in hours:minutes:seconds and bars (MEAS) as well as beats. This illustrates the musical relationship between the Project tempo and time signature (→Tempo Map).

Fig.: SMPTE time and measure display

Meter Bridge (MB-24)

After moving the cursor to the MEAS or BEAT field, the →Data Dial can be used to move the project in bar or beat steps.

Tip

- Numeric entry allows the measure and beat to be specified (→NUMERICS, not on the VS-2000).
- The first beat can be anywhere within the Project (→Sync Offset Time).

Meter Bridge (MB-24)
→Ext Level Meter (MB-24) (p. 72)

Metronome
The VS's internal metronome gives the recording a specified tempo. The tempo value can be constant or variable (to accommodate tempo changes). To edit track data based on the →Grid, it is recommended to activate this "click" for all material recorded. The M. provides several click sounds for strong and weak beats as well as a simple programmable rhythm computer*. While the tempo and time signature need to be set in the →Tempo Map, global settings are to be found on the "Metronome" page:

Menu:	Utility → Metronome
VGA:	SHIFT + click
VS-2000:	Rhythm Track → Metronome

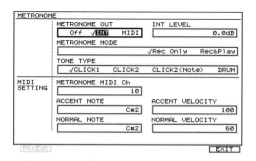

Fig.: Metronome settings

Use "INT" in the "Metronome Out" section, or click * to activate the metronome.

[1] The VS-2000 provides a Rhythm Track that is, in fact, a complete rhythm composer.

*[2] Yellow icon: internal metronome sound
Blue icon: MIDI notes are transmitted*

To hear the metronome during playback, set "Metronome Mode" to "Rec&Play". The various sound combinations for strong and weak beats can be selected in the "Tone Type" field.

Note:

- The metronome level can only be set in the *Int Level* field – not via MONITOR:

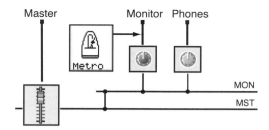

Fig.: Metronome level

- When →Sync Offset Time is active, the metronome starts at the position that corresponds to the selected offset.
- With the exception of the VS-2000*, the metronome sound is only sent to the →Monitor Bus. This has the advantage that the metronome is audible but can't be recorded by accident.

Tempo And Time Signature

These parameters need to be set in the →Tempo Map:

Menu:	Utility → Tempo Map
VS:	SHIFT + TAP MARKER
VGA:	SHIFT + click TAP
VS-2000:	Rhythm Track →Tempo Map

Rhythm Computer:

The "DRUM" parameter in the "Tone Type" field allows offers a simple →Rhythm comput-

Microphone Modeling

er. The VS-2000, on the other hand, provides a rhythm composer function in the form of a →Rhythm Track.

Recording The Metronome Sound
When activated as source for the →Generator/Oscillator, the metronome can be recorded to a user-defined track using the →Bouncing operation. This is the way to record the sound of the internal →Rhythm computer particularly for the VS-24xx.

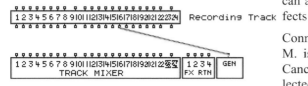

Fig.: Routing the generator to a track

Sending The Metronome Signal
The M. can be sent out of the VS via an →AUX Bus or, more effectively, via a →Direct Path. To do this, deactivate MIX in the →Generator/Oscillator menu. Click a DIR field (DIR 1, for example) or set an AUX knob after programming the appropriate →Output Assignment to route the metronome signal to the desired output.

Mickey Mouse Effect
This effect is caused by raising the pitch of recorded signals, which also shifts the →Formants. To create it in real-time, select →Pitch Shifter or →Voice Transformer →Effect Patch, or increase the playback speed (→Vary Pitch).

MicModeling
→Microphone Modeling

Micro Edit
→Automix (p. 19)

Microphone Modeling
M. is an →Effect that simulates various microphone sounds. It requires the use of one of the supported reference microphones. In some instances, it can even be used to add a microphone impression to a line signal.

Microphone modeling

Effect patch:	P110~138
Connection:	Insert
Application:	Useful

Even though the characteristics of the microphone models are only impressions of the originals, M. nevertheless enhances the sound of signals recorded with budget microphones. It can also be used for interesting coloration effects

Connection:
M. is a dual-mono effect. To avoid →Phase Cancellations, "InsL" or "InsR" needs to be selected when using it as an →Insert effect. This allows two separate signals to be processed.

Settings
"CnvA" and "CnvB" are the actual converters for the two channel signals. Use "Input" to specify the microphone used at the time of recording (Reference).

Reference Microphone	Name
Roland DR-20	DR-20
Shure SM 57	Sml.Dy
Dynamic headset	Head.Dy
Miniature condenser	Min.Cn
Linear	Flat
AKG C3000B	C3000

"Output" allows the selected microphone sound to be modeled:

Simulated microphone	Name
Shure SM 57	Sml.Dy
Shure SM 58	Voc.Dy
Sennheiser MD 421	Lrg.Dy
AKG C 451	Sml.Cn
Neumann U87	Lrg.Cn
Neumann U47 (tube)	Lrg.Cn
Linear	Flat

Feel free to experiment with source microphones not mentioned in the list – the result

Microphone Level

may be surprisingly good.

The "Dstn" (distance) block simulates the proximity effect and delays caused by varying the distance between the source and the microphone ("Time", in cm).

Tip

"Mic Modeling" can be used as channel →Delay to compensate for offsets between various microphones (i.e. main and auxiliary/ambient mics). To do so, deactivate all blocks that are not directly instrumental in achieving this effect. The value is specified in [cm], so that no additional conversions are necessary.

Microphone Level

The low voltage of microphone signals is usually 2mV for dynamic and 20mV for condenser microphones. Such signals need to be boosted by up to 60dB (x 1000) to achieve a proper working level. This requires a good →Microphone Preamp in order to maintain the signal's integrity.

Microphone Preamp

→Microphone Levels* need to be adapted to the console's working level, requiring extremely high level boosts (often with a factor of 1000, i.e. 60dB). The single most important criterion for the M.'s quality is its noise floor. For critical applications, using an external high-end microphone preamp may be necessary. The VS's →Input Mixer section provides the required M.s. The necessary signal levels can be set with the GAIN knobs (SENS).

* –30 ~ –50dBu,
 for soft signals up to –80dBu

The VS-2480 provides a pad switch to attenuate incoming signals by 20dB. This is only necessary for some →Line Level signals. If an external preamp is to be used, choose a model with a digital output (→S/PDIF, 24 bits) to avoid unnecessary D/A and →A/D Conversions. Preamp models, fitted with an analog (!) compressor/limiter are recommended. It can be used to reduce the →Dynamic range of the source material before the A/D conversion, enabling work at higher levels (→Level Setting While Recording) on the VS. (See also →ADA-7000 (p. 13) and →Preamp Modeling (p. 163)).

MIDI

M. is a standardized interface for the transmission of data between musical instruments and computer-controlled effect devices, mixing consoles, DAWs, etc. Sounds or audio data are not transmitted in this procedure, only control information, which goes from a transmitter (Tx) to one or more receivers (Rx). The VS provides MIDI support for the following functions:

1. Recording the VS mixer settings with a sequencer:
   →MIDI Mixer Settings (p. 124).
2. Selecting effect patches from an external device:
   →MIDI Effect Patch Select (p. 124).
3. Sending and receiving effect patches and Scenes:
   →MIDI Bulk Dump (p. 123).
4. Sending and receiving timecode and transport commands during a synchronized operation: →Synchronization (p. 217).

MIDI data is sent ("Tx") from a MIDI OUT socket and received ("Rx") if that socket is connected to the external device's MIDI IN port. The "MIDI" menu allows the necessary parameters for the MIDI functions to be activated:

Utility → MIDI Parameter

Tip:

- Received Midi data, is displayed in the VS's MIDI indicator.
- To check whether the VS also sends MIDI messages, move any fader up and down. This only works, however, if the →Mixer Control Type parameter is set to "C.C."

MIDI Bulk Dump

→Scenes, User →Effect Patch and →EZ Routing User Templates of the current Project can be transmitted to a MIDI sequencer or a second VS (of the same model), and can be sent back again. This is the only way to transfer Scenes from one Project to another Project* or VS.

* *To transfer Scenes form the current to a new Project, use →Copy Mixer Parameter.*

1. Transmission (Bulk Tx)

Establish the required →MIDI Connections, then carry out the following settings. Scenes, EZ Routing templates and effect patches have separate "Bulk Tx Sw" switches to be activated for transmitting. Set the MIDI sequencer to record and press [YES] to start sending the selected data.

Utility → Midi Parameters 1:

SysEx.Tx Sw = On → F5 BlkDmp →
BULK Tx Sw = On → F1 BULKTx → YES

2. Receive (Bulk Rx)

Establish the required →MIDI Connections, then carry out the following settings: The VS automatically recognizes the data type (Scenes, EZ Routings or effect patches) being received.

Note:
Be aware that the VS stores data received in this way in their original memory locations. Don't forget to copy important data to other memories before sending bulk data.

Utility → Midi Parameters 1:

SysEx.Rx Sw = On → F5 BlkDmp →
F2 BULKRx wait for a message
→ YES →…Applied? → YES

Device ID:

The →Device ID (address of a given MIDI device) is contained in the bulk data. Reception only works if the receiver uses that address.

Tip:
If two operators work on the same Project, but on separate VS machines and in different locations, the mixer settings (Scenes) can be exported as →SMF data and sent via e-mail. This has the advantage that the audio material (usually too big to send via the internet) only needs to be burned to CD-R and sent once.

MIDI Clock

→MIDI Sequencers or a drum machines can often only be synchronized via. M. Their 1/96th-note resolution is far less precise than →MTC, but it includes all tempo information of the song as well as the transport commands (Song Position Pointer). The VS also uses the "MIDI Clock" format to transmit the VS's →Tempo Map and →Sync Track information. No additional settings are necessary to use it. Activate the "MIDI Clock" parameter to ensure that the VS sends M. data.

Utility → Sync Parameter:
MIDI OUT SYNC Gen. = MIDIclk (or SyncTrk)

Whereas the VS can transmit MIDI Clock information, it can only be synchronized by receiving MTC signals.

MIDI Connections

Depending on the use, →MIDI communication only works if the following connections are established:

1. Sending data from the VS

Connect the MIDI OUT socket to the MIDI IN port of the second VS or another device.

VS	MIDI Sequencer
MIDI OUT	MIDI IN

Check whether the "MIDI" menu uses the following (default) setting:

Utility → MIDI Parameter →
"MIDI OUT/THRU"= OUT

MIDI Effect Patch Select

2. Receive
Connect the VS's MIDI IN socket to the MIDI OUT port of the second VS or another device.

Midi sequencer	VS
MIDI OUT	MIDI IN

In most instances, one cable connection is enough for sending data one way. For two-way communication, be sure to set the →Control Local Sw to "Off" to avoid MIDI loops.

MIDI Effect Patch Select

The VS's internal effect patches can be selected from an external device via →MIDI →Program Changes. This only works if the VS uses the following setting:

Utility → MIDI Parameter →
"EFFECT P.C. Rx Sw"= On

The effect processors 1~8 correspond with the same numbered midi channels (channel 1: effect processor 1, etc.).
Whereas Presets 0~99 can be selected with program change messages only, the other preset patches require message clusters consisting of additional bank select messages.

Presets: MSB = 0   LSB = 0 - 2   P.Chg.xx
User: MSB = 0   LSB = 3 - 4   P.Chg.xx

The following example (taken from "Logic") shows how to select patch 207 for processor 2.

1	1	1	1 Control2	0	0 Bank MSB
1	1	1	2 Control2	32	2 Bank LSB
1	1	1	3 Program2	-	7 Clavinet

Fig.: Bank select messages in Logic

MIDI Mixer Settings

The VS's mixer settings can be controlled via MIDI. These can be recorded, archived, edited and played back on an external sequencer, in which case the VS reproduces them. This is actually more flexible than working with the VS's →Automix.
Remote control is possible with →Control Changes (easier to grasp, yet less flexible) or →System Exclusive Data.

Prm	C.C	Excl	Prm	C.C	Excl
Fader	x	x	CH Link	-	x
Pan	x	x	Phase	-	x
EQ	x	x	DIR	-	x
EQ Sw	x	x	FX Ins	-	x
EQ Res	-	x	Solo	-	x
Dyn	-	x	Mute	-	x
A.-Send	x	x	Tr Status	x	x
Aux Sw	-	x	EZ Rout.	-	x

After establishing the →MIDI Connections between the two devices, the following settings in the "MIDI Parameter" menu (Utility) need to be made:

Remote Control Using Control Changes:

Control Local Sw: Off
Mixer Control Type   C.C

Remote control using SysEx Messages:

Control Local Sw: Off
Mixer Control Type   Excl
SysEx Rx   On
SysEx Tx   On

Any setting change performed on the VS, in accordance with the tables shown above, is transmitted from its MIDI OUT socket to a synchronized sequencer (→Synchronization). When played back, the data the VS receives via its MIDI IN socket will have identically the same parameter changes.

MIDI OUT Sync Generation

If the VS acts as →Synchronization master, this parameter can be set to →MTC (p. 133), →MIDI Clock (p. 123) or →Sync Track (p. 215) (depending on what the Slave requires).

MIDI Scene Change

The VS's →Scenes can be selected from an external device via →MIDI →Program Changes. For this to work, the following settings need to be carried out in the VS:

> Utility → MIDI Parameter →
> "SCENE P.C. Rx Sw"= On

Sending program change numbers 0~99 to channel 16 cause respective Scene changes when in Play mode.

Note:
If the VS receives a "Program Change" message during playback, it stops, selects the Scene, and then resumes playback. While recording, the VS ignores all MIDI messages.

MIDI Sequencer

M.s are used to record, edit and play back MIDI data. They come either as independent devices ("hardware sequencers") or as software (Cubase, Logic, Cacewalk etc.). They can be used with the VS for a synchronized Midi and Audio playback (→Synchronization), to transmit mixer data and to automate all track and mixer parameters. The VS can also be practically used as a →Control Surface for the ergonomic operation of software sequencers.

Mix

Setting the level balance and sound/timbre of all required audio data and recording the result to a stereo track* is called "mixing". This is the same for playback tracks, →Effect Return and signals received by the →Input Mixer (sound synchronized midi sound modules, external effects, etc.) (which might be the audio of MIDI modules controlled from a synchronized sequencer).
On the VS, mix operations can be simplified using Scenes, the Locator, intermediary mixes, recording effects to V-Tracks, automation, etc.

> * →*Surround mixes require 6 busses and as many master tracks.*

1. Preparation
Start by having a conceptual idea about the kinds of sound desired. Carefully listen to the material and decide whether all tracks are to be used at all times in the arrangement ("less is more"). Check how many effects processors there are, name the playback tracks (→Name Entry), prepare a reference CD (→A/B Comparison), etc. Clean up the audio material by removing background noises or passages that shouldn't be there (→Phrase Trim, →Region Erase), and set the right crossfades (→Phrase Parameter). An entire song may even need to be copied to a different location for alternative arrangements (→Region Copy, →Region Arrange). Due to the →Effect Economy it is possible to record effects returns to use the processor several times (→Effects, recording). To quickly jump to certain song locations, a →Foot Switch can be used (define Markers, set →PREVIOUS/NEXT to "MARKER").

2. Mixing
The actual mixing process not only requires technical abilities but also clear ideas about which sound to create for the Project at hand. It is therefore recommended to make an →A/B Comparison with a reference CD to help guide certain choices. Such comparisons often minimize listening fatigue. Furthermore, be sure to use different listening levels. Viewing interim results by comparing before and after sounds is enabled via →Scenes.

Level, Pan
Firstly, all unfiltered signals should be leveled in the mix according to their importance. Watch out for masking effects. The vocal part often appears louder when the level of instruments that occupy the same frequency range is reduced. The kick drum, bass, solo vocal and solo instruments should be located at the center of the stereo image. Instruments with similar frequency ranges can be assigned to opposite channels (HH right, shaker left, etc.). The stereo balance must be checked primarily by listening to the result and only for control purposes by watching the bus level meters.

Mix

Equalizer

Transparent mixes are based on avoiding frequency overlaps and masking effects. This is done by creating "frequency windows" with the channel EQs. Ideally, each instrument or voice gets its own frequency range. One important step is to use the →Low filter (or →High-Pass) to cut the 100~200Hz range for all signals except the bass and the kick drum. Other frequency ranges can be "cleaned" in a similar way to ensure that the most important sources in those ranges are clearly audible. High-frequency boosts should be restricted to the main vocals, to make them more conspicuous. Conversely, consider lowering the →High band above 8kHz for signals that should be "at the back".

Tip

The VS-2480's →Bandpass filter (BPF) can be used to test spectral signal distribution by means of frequency windows. Assign the center frequency parameter to the "Pan" knobs (→Knob/Fader Assign) or faders (the fader positions then indicates the frequency) for comfortable work.

Dynamic Processing

Only the VS-2480 provides both a →Compressor and an →Expander effect for all channels. The remaining models should use the compressor, for dynamics applications, while background noises need to be eliminated by →Editing.

It is recommended to slightly compress all individual channels rather than using heavy compression while →Mastering for the Mix Track. The VS's channel compressors are ideally suited for this approach. The default settings often yield excellent results. It is possible, however, to raise the Ratio value to achieve a gain reduction (GR) of up to 6dB (→Dynamics).

Effects

→Reverberation is the most important →Effect for any mix. To enhance the depth impression of the sound image, consider using two or three different reverb types. This can be achieved even without additional →Effect Boards, by recording effect signals (→Effect Economy). Use a bright reverb with a PreDly setting of 60~120ms for the more up front signals (vocals, solo instruments). Use rounder reverb patches with longer reverb times to add depth to the remaining instrumental signals. Depending on the result desired, patches P036~P077 usually provide a good starting point.

A →Delay (P078~P090) synchronized to the song tempo fattens the sound and creates a different depth sensation for sustained notes. The delay's →Effect Return could be fed to a reverb processor. →Chorus effects are often used to "broaden" relatively dull signals and add some life. Among the other preset effects, the advantageous use of "Chorus RSS" (P009) is to be emphasized. Use the →Microphone Modeling effect to add a "microphone sound" to line signals. The "Time" parameter even allows such signals to be delayed (set in [cm] units). Special effects like →Vocoder, →Voice Transformer, →GuitarMulti, etc., should be recorded before mixing begins.

Fader Groups

→Fader Groups (not on the VS-2000) allow the levels of several instruments to be set simultaneously. For example; assign all drum tracks (TRACK MIXER) and percussion parts of a synchronized MIDI module (INPUT MIXER) to the same group to be able to set "the rhythm section" louder and softer with just one fader.

Depth Of The Sound Image

The transparency of the mix not only depends on the frequency distribution, but also on the various levels of instruments in the mix. This can be achieved by carefully setting the levels, slightly reducing the high frequencies (only the high range of vocal or solo instrument parts should be boosted), and appropriate reverb programs.

3. Recording the Mix

The final mix is normally recorded to two empty →V-Tracks in the Mastering Room.

Mix

VGA: CD-RW Mastering → Mastering Room
VS: SHIFT +CD-RW/MASTERING

Here, as in the Mastering operation, the mix will be recorded onto two "Mix tracks" (→Master Tracks). If a subsequent Mastering is intended, the →CDR parameter must be set to „Off". This operation is carried out in the same way as is described in →Mastering Room (p. 115).

Brief Instructions

- Activate the Mastering Room, set "Status" to "Record".

VGA: Click the REC icon [MSTR ●]
VS: Activate with the dial; STATUS = "REC"

The CDR-W/Mastering button flashes. In this mode, all tracks can be played back and mixed.

- Select the V-Track pair to be used as →Master Tracks (stereo). There are 16 V-Tracks available for as many different independent mixes.

- After Record

  The position of the "Master" tracks can be freely selected.

  In most instances, the "Zero" default setting can be used. Select "Stay Here" for corrections to the mix (p. 127).

After Record	Position M.Tr.
to ZERO	00:00:00:00:00
to last Phrs: 0 s / 2 s / 4 s	Behind the last Master Track
stay HERE	Same time position like other tracks

→Master Effect shouldn't be used while committing a mix to two V-Tracks. Use the processing power for other, more important, applications. Furthermore, →Mastering should be a separate process, because it requires an undivided attention. By recording the mix to regular tracks (not in →CDR mode!), editing the mix later becomes an option.

- Press REC + PLAY to start recording the Mix tracks (Master Tracks). All settings made and all →Automix events activated, affect the tracks being mixed. It is possible to leave the "Mastering Room" if necessary to set mix-related parameters in other menus (e.g. →TR F/P, →CH EDIT, etc.). Use the MASTER fader for fades.

3. Playing Back The Mix Tracks

3.1. The Mastering Room
Select the "Play" status to hear only the master tracks during playback. The INPUT MIXER and TRACK MIXER channels are muted in this mode (→Fig.: Master tracks during playback (p. 116)). Playback and →Positioning work as usual.

3.2. As Project Tracks
Master tracks *not recorded in CDR mode* are considered regular Project →V-Tracks. After deactivating the Mastering Room and selecting the V-Tracks (usually V-Track 16), the mix tracks can be played back via tracks 23/24 (linked for stereo operation). Use →Solo mode or mute all other tracks.

VS-2000:
Use →Region Move or →Phrase Move to move the master tracks to free tracks (→Phrase Move allows them to be dragged to the desired tracks). The mix tracks can then be edited by →Mastering them.

4. Correcting A Mix Without Redoing It
If the overall mix seems acceptable, whilst the main vocals or certain instrument levels require adjustment, just that aspect can be corrected. Given that the mix is a stereo track, which is perfectly in phase with the original tracks, the track that should be louder can be added, resulting in raising its level.

Tip
It is also possible to reduce the level of specific signals. As the signals to be corrected are available in two versions (mix and separate tracks), reversing the phase of an individual

Mix Bus

track cancels out that signal. To do so, use the →Phase switch to shift the phase by 180° (Inv) and set the level again.

Mix Bus

For monitoring and mixing purposes, all mixer and →Effect Return channels are routed to the stereo Mix (MASTER) bus. They are all firstly sent to the Master →Inserts first (if necessary), and from there to the →Master Fader and the MASTER OUTs.

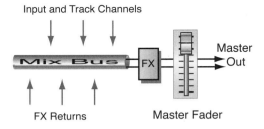

Fig.: Mix bus

The →MIX Switch available on all channels allows channels to be disconnected from the M. (which may be necessary to connect →External Effect Processors). Use →Output Assign to route optional analog and digital signals to the mix bus (by default, the Monitor bus is sourced from the M.).

MIX Switch

In most situations, the mixer channels are routed to the →Mix Bus. During →Recording, however, the VS temporarily disconnects input channels that are routed to the tracks selected for recording*. Special applications, like inserting →External Effect Processors or monitoring only the effect depth of individual signals for a →Send/Return Effect, also require deactivating the Mix switch.

This can be done in the mixer channel's "CH Edit" menu:

    CH EDIT   Click [MIX] .

* *These input channels are monitored via the TRACK MIXER.*

Mixer

1. General Considerations

A mixer (mixing console) combines several signals, whose levels, sounds and Pan can be edited, to the →Mix Bus and other busses. It transmits the resulting signals to a master device for the monitoring system, to additional effects devices, etc. Each incoming signal is assigned to a separate →Channel Strip used to change the signal's level, Pan setting, equalizing, dynamics, and even output assignment. While analog M.s provide dedicated controls for every parameter of a channel, digital mixing consoles provide only a limited number. Their functions depend on the currently selected mode or need to be assigned via the display (→Fader Buttons, →Function Buttons). Working with a mouse and a VGA screen greatly simplifies the operation, while also providing graphic representations of the equalizer and dynamics settings.

2. Mixer Menus

Mixer Menus contain several parameters for specific channel groups (mixer sections). They can be selected from →Mix or →Automix.

Tip

The fader settings of the INPUT MIXER and TRACK MIXER levels can be set simultaneously: Select one section on the VS (and watch the physical faders) and assign the other to the display (PAGE 2/3: →IN F/P or →TR F/P).

2.1 VGA screen

The VGA screen's →HOME page allows the following mixer pages to be selected from →PAGE 2/3:

- →IN F/P (p. 96) and →TR F/P (p. 226) "Input or Track Mixer Fader/Pan"
  Easy access and editing of faders/Pan/ AUX Sends and activating EQ, dynamics, Mute, Solo, and Automix.

- →MltChV (p. 130) (not in the VS-2000) This page provides an additional graphic representation of the EQ and dynamics settings.

- →V.Fader (p. 235):
In this mode, the faders can be configured for remote control of a →MIDI Sequencer and/or sound modules, etc. The VS-2480's Pan knobs can also be used as →Control Surface.

2.2. LCD

Due to its size, the LCD cannot produce a complete display of all faders and Pan knobs. One of the following sections must therefore be selected: Input Mixer, Track Mixer, AUX + Direct Master + FX Returns. These parameters can also be edited with the mouse. Press PAGE to assign the →Function Buttons the "IN F/P • TR F/P • AUXF/P • …" functions.

- IN F/P: Input Mixer, Fader/Pan
- TR F/P: Track Mixer, Fader/Pan
- AUXF/P: Fader/Pan positions of the →Aux Master, →Direct Path masters and the →FX Return Channels

3. Input Mixer and Track Mixer functions

The VS contains two mixers used for different applications. Understanding their functions is of the utmost importance for working with the VS recorder. Before editing a mix parameter, the appropriate MIXER level needs to be selected with the →Fader Buttons or via the "Mixer" menu.

3.1. Input Mixer

The INPUT MIXER routes incoming signals either to recording tracks (for recording) or (and) to the mix bus for listening and mixing. The number of INPUT MIXER channels depends on the VS model:
VS-2480 = 24, VS-2400 = 16, VS-2000 = 10.
To take advantage of all supported channels, extra hardware must be used (→AE-7000, →AES/EBU). it's not possible in the VS-2000).

VS-2000:

The INPUT MIXER channels cannot be set using the physical faders. Instead, for editing purposes, one of the (round) INPUT buttons 1~10 needs to be pressed. This limitation doesn't apply to the VGA display.

3.1.1. Recording

During recording, incoming signals are affected by the INPUT MIXER settings. These channels need to be routed to the recording tracks of the →Hard Disk Recorder (→EZ Routing).

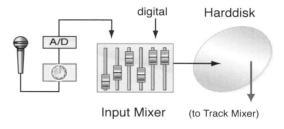

Fig.: Input Mixer during recording

Unable to be seen by the user, channels routed to record-ready tracks are disconnected from the mix bus to avoid such signals being outputted in duplicate (via the INPUT and TRACK MIXER sections). Level and tone settings performed by the INPUT MIXER affect the signals being recorded (e.g. →Equalizer and →Insert effects). The signals heard in these situations come from the hard disk and are sent to TRACK MIXER channels, where they are considered playback channels (see Fig.: Input Mixer during recording).

3.1.2. Playback

During playback, and for all INPUT MIXER channels not routed to armed tracks, the incoming signals are automatically transmitted to the →Mix Bus. During playback, there is no difference between the INPUT MIXER and TRACK MIXER functions (see figure).

3.2. Track Mixer

The TRACK MIXER channels are hard-wired to the corresponding HD tracks. Channel 1 is linked to track 1, channel 2 to track 2, channel 24 to track 24 etc. This channels are known from a conventional studio as "Tape Returns". The track signals are usually transmitted to the →Mix Bus (see Fig.: Input and Track Mixer during playback).

Mixer Control Type

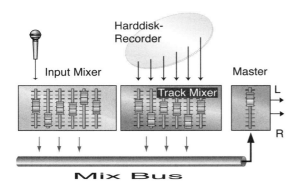

Fig.: Input and Track Mixer during playback

3.3. Effect Return Channels
The output signals of the effect processors are connected to →FX Return Channels and transmitted to the mix bus. Such channel strips have a simpler structure: they don't provide EQ, dynamics, insert, etc., parameters.

3.4. Channel Strip (Channel Edit)
The mix, →Equalizer, and →Dynamics parameters of a channel need to be set via the →CH EDIT (p. 42) menu.

3.5. Master Edit Section
The →Master Section (p. 112) is used to set the level and stereo balance of the →Mix Bus, the Monitor Bus and the →Aux Master/→Direct Paths

Mixer Control Type
The M. parameter allows the Midi →System Exclusive Data or →Control Changes to be selected for the transmission and reception of →MIDI Mixer Settings.

Utility → MIDI Parameter → "Mixer Control Type"

MltChV
The "Multi Channel View" VGA page (not on the VS-2000) provides an additional graphic representation of the EQ, dynamics and surround settings of all mixer channels. This display is even clearer than anything one might expect from an analog →Mixer.

VGA: PAGE 2/3 → F4 MltChV
Menu: Mixer → MULTI CH STRIP VIEW

Use the VS's →Fader Buttons or the tabs in the right half of the display to select the desired mixer section.

EQs and Dynamics:
Use the display fields to switch these on or off. The VS-2480 also has a →Channel Strip, which can be assigned to the Pan knobs by pressing a button or clicking the CH field (top left).

Surround:
While →Surround mode is active, the channels' surround position can be set using the mouse.

AUX Sends:
Click an AUX field to select the cursor icon, then drag the mouse up or down to set the AUX Send level.

V-Track:
Click a V-Track field (beneath the Track number) to open the assigned pop-up menu and select a V-Track. Only the main track is shown here.

Link:
Click a CH field below a fader to activate Fader or →Channel Links.

Effects:
Clicking an effect icon in the AUX/FX section (VS-2480: TR 17–24) opens the corresponding effect menu for choosing an →Effect Patch (if not, the effect is routed as an insert). The assignments of AUX busses to the effect processors can also be changed here.

The VS's LCD is also able to display one parameter at a time for several channels. The →PRM.V enables parameters to be edited.

MMC
"MIDI Machine Control" data transmits transport commands and the current time value

Monitor Mix

from the master to the slave. Rather than →Timecode data, these messages are merely system control data in the form of →SysEx data. During →Synchronization via →MTC, MMC needs to be activated separately for the master and slave:

Utility → Midi Parameter:

Midi Parameter	Master	Slave
MMC Mode	MASTER	SLAVE
SysEx Sw	TX = ON	RX = ON

Monitor
I. "Monitor" refers to an additional listening mix besides the Main Mix (e.g. →Headphone Mix). The VS transmits these signals to the →Monitor Bus, and from there to the →MONITOR OUT sockets.

II. The term also refers to the speakers used in a studio (→Studio Monitors) or on stage.

Monitor Bus
The VS's M. is an additional stereo →Bus used to independently listen to the signals of various busses or paths. The M. is usually sourced from the →Mix Bus (MSTL/R). For control purposes, however, it can be routed to other busses (→Output Assign):

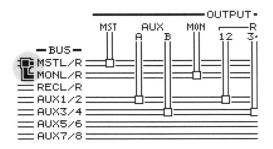

Fig.: Assigning the Monitor bus on the VS-2400 (LCD)

On the VS-2400, the headphones are hard-wired to the Monitor bus. The other models allow specific busses to be assigned to the headphones (see the tip).

Routing On The Monitor Bus (VGA):
E.g. the coaxial Out to feed digital Monitors.

1. Select the desired output socket.
2. Click the patch nodes of the currently selected bus and hold down the mouse button.
3. Drag the mouse up or down to the Monitor bus and release the button*.

   * Alternatively, click a patch node and change the assignment using the dial.

Routing The Monitor Bus (LCD):

1. Use the cursor buttons ◀▶ to select the desired output.
2. Via dial select the desired connection.

Tip:
- Speaker Level Via MONITOR Knob:
  If the speakers are connected to the MONITOR sockets, the listening level can be set with the MONITOR knob. If digital monitors are used, connect them to the optical or digital socket.
- Separating headphones from Monitors:
  Route PHONES to MASTER and set the listening level independently of the Monitor bus with the PHONES knob (not on the VS-2400).

Metronome
With the exception of the VS-2000, the metronome sound is only sent to the →Monitor Bus. This allows the metronome to be used even when transmitting signals to a PA system or while recording the →Master Tracks: it won't be heard in the wrong places, or recorded.

Metronome Level:
The metronome signal is inserted after the MONITOR knob. Consequently, the only way to change the metronome's level is by using the "Int Level" parameter (→Fig.: Metronome level (p. 120)).

Monitor Mix
This is another term for the →Headphone Mix (studio) or →Live Mix (for the musicians on

stage). The M. is played back using separate speakers. In most instances, there will be several such mixes, which can be prepared on the VS by using its →AUX Busses. Individual signals (e.g. the click track for a drummer) can also be transmitted from a →Direct Path.

MONITOR OUT

The two M. sockets usually feed the →Studio Monitors and the M. level is determined by the MONITOR knob. The MASTER fader, however, is used to set the overall mix bus level. These →Outputs are balanced and have a impedance of 600 Ω.

Mono

Signals recorded with only one microphone, or transmitted from one output are "monaural" (from the Greek, often called "mono") and only require one mixer channel. The channel's →Panorama knob nevertheless allows mono signals to be positioned anywhere in the stereo sound image (between the left and right speakers). The VS's →FX Return Channels provide a MONO switch, which can be used to combine the effect's →Stereo channels to a mono signal for special effects.

Mono Compatibility

Because produced mixes are very likely to be played on mono devices (transistor radios, TV sets, etc.), the stereo mix should also sound almost in mono (i.e. be mono compatible). Such mono signals are usually a combination of the left and right channels. In particular signals with opposite phases, which create no apparent problems in stereo, may cancel each other out or be affected by the →Comb Filter Effect in mono. Stereo effects derived from phase shifts (→Chorus and →Flanger) as well as →Reverberation appear weaker and a lot less spectacular in mono. Also bear in mind that signals at the center of the stereo mix appear louder when played back in mono. The VS's →Master Section provides no mono switch which means that a listening device must be used to hear the product of a mono sound.

Tip
- The "Small Radio" or "Small TV" →Speaker Modeling algorithm creates a mono impression of the →Mix.
- Adding a mono reverb signal (→FX Return Channel) to the stereo mix enhances the effect when played back in mono.
- The best results are obtained with an additional mono mix.

Motorized Faders

The VS-24xx models are fitted with M. This has the advantage that the faders always assume their correct positions when →Mixer sections, recall →Projects or →Scenes are switched, or the →Automix function used.

Mouse

The PS/2 mouse must be connected to the MOUSE socket on the VS's rear panel (the VS-2000 needs to be fitted with an optional VS-20 VGA board). There are several parameters that allow the mouse's responses to be set according to preference.

Utility → System3 → *PS/2 Mouse*

Here the mouse can be activated/deactivated, the pointer speed set and the functions of the left and right mouse buttons reversed for left-handed users.

Mouse Editing

The →Mouse* can only be used after activating the mouse operation:

Utility → System Parameter3 → PS/2 Mouse

Mouse editing is only possible in the →Operation Display. Switching between the LCD and the VGA screen is performed as follows:

VGA:	Utility → System Parameter2 → VGA Out
VS-2480:	SHIFT + UTILITY
VS-2400:	SHIFT + MASTER EDIT
VS-2000:	HOME + F6

Multitrack Recording

Region For Editing
Draw a frame in the desired time and track regions to specify the area to be edited (→Region Edit and →Automix edit). Click anywhere outside the Region to cancel it.
To move the Region, position the →Cursor above it (the 🖐 icon appears). Click and hold the mouse button. The mouse icon changes to 🖐 the "hand closed" icon. Additionally pressing the SHIFT button adds a "+" to the mouse icon. Move the icon to its destination area and release the mouse button to execute the operation.

Phrases
To edit a →Phrase, click anywhere in the track bar area. That position becomes the →FROM point (→Edit Points). The operation for moving the Phrase is the same as for moving Regions.

Entering Values
Click a parameter. The cursor icon changes to ↕ (upward/downward pointing arrow). Drag the mouse up or down while holding down the mouse button to change the parameter's value. Dragging the mouse to the left or right decreases/increases the value in finer steps. Holding down SHIFT while dragging increases or decreases the step size (this depends on the selected parameter).

Right-Clicking
Right-clicking the →Playlist opens the following edit menus: →Zoom (p. 245), →Wave (p. 241), →Scrub (p. 202), edit operations, →Undo (p. 231), →Grid (p. 86) and →Edit Message (p. 60). Right-clicking the →Measure Bar calls up the menu for →Edit Points (p. 61), →Punch In/Out (p. 170), →Loop (p. 108) and →Marker (p. 109).

Move
→Region Move (p. 191), →Phrase Move (p. 152), →Automix (p. 19)

MT1, MT2
→Record Mode

MTC
MIDI Time Code (MTC) is used to ensure →Synchronization of two devices via →MIDI. Based on the →SMPTE format, it is far more precise and flexible than →MIDI Clock. It contains only the absolute time position and no song tempo information:

Hours: Minutes: Seconds: Frames: Subframes

Synchronized operation only works as expected when both the master and the slave device use the same →Frame Rate. MTC is used in combination with the transport controls →MMC.

MTK
→Mastering Tool Kit (p. 117)

MTP
With the exception of the VS-2000, this is the recommended →Record Mode (24 bits, compressed) for →Project New.

Multi Channel View
Only available in the VGA of the VS-24xx models, a variety of parameters for 12~16 mixer channels can be simultaneously displayed (→MltChV).

Multiband Compressor
→Mastering Tool Kit (p. 117)

MultiTap-Dly
This effect contains 10 →Delay lines whose delay time, level and stereo placement can be varied independently. This can be used for experimental effects.

Multi Tap Delay	
Effect patch:	P085~P086
Connection:	Send/Return
Application:	Experimental

Multitrack Recording
Multitrack recordings can be created by re-

cording all parts simultaneously (but to separate tracks) or by recording one track after another.

Mute

A console's Mute function allows chosen channels to be silenced. For control purposes, this therefore lets certain channels be left out of playback.

VGA:	Click the M icon in the →Playlist, →TR F/P or →MltChV.
CH EDIT	Click MUTE
VS-2400:	MUTE → press a flashing CH EDIT button
VS-2480/ VS-2000	SHIFT + MUTE → press a flashing CH EDIT button

Muted channels are flagged with a "MUTE" message in the footer row (yellow) of the VGA screen, or in the header row of the LCD.

Switching Off Mutes Individually

Even after deactivating MUTE mode by pressing the VS's button of the same name, muted channels preserve their status. This is indicated by means of a flashing "MUTE" message.
Muting for each channel needs to be deactivated individually. On the VGA screen, this is done by clicking the yellow icons.

Switching Off All Mutes

All muted channels can be deactivated using a convenient shortcut. Be aware, however, that Mute mode as such remains active, even though all mixer channels are once again audible. Here is how to return to normal operation (in which case the VS's MUTE indicator goes dark).

VGA:	Playlist: click the arrow at the bottom of the mute column
VS:	CLEAR + MUTE → Deactivate Mute mode (see above)

The →Solo/Mute Type parameter allows the restriction of the Mute function's influence to the mix bus.

Tip:
- →Parameter View is also used in the LCD to control the Mute settings for 24 INPUT MIXER or TRACK MIXER channels simultaneously.
- If necessary, the influence of the Mute function can, via the →Solo/Mute Type parameter, be restricted to the mix bus. Otherwise the Mute function applies to all busses. Remember this when preparing a Monitor mix for the musicians.

Automix

Mutes can be integrated into the →Automix.

VS-2000 + LCD

It is possible to simultaneously activate or deactivate the mutes of all mixer channels in the LCD. Move the cursor to the MUTE field on any CH EDIT page. Hold down CLEAR + rotate the dial to select the desired status for all chosen channels.

N

Name Entry

The VS allows a name to be entered for the following functions and parameters: →Project (p. 165), →Track (p. 226), →Phrase (p. 147), →Take (p. 221), →Locator (p. 106), →Marker (p. 109), →Scene (p. 200), →Routing (p. 198). Naming is performed in the following menu:

Fig.: Entering names (example: track)

It may not be necessary to use all of them, but

naming the Project and playback tracks with an expressive name is recommended. Characters can be entered in various ways. The F buttons' functions are always the same:

F3 (Delete):
Erases the character displayed in reverse.

F4 (Insert):
Pushes the character one position to the right.

F2 (Backspace):
Moves subsequent characters to the left and erases the selected character.

History:
F1 allows all names entered since switching on the VS to be consecutively recalled. This function may save a lot of time.

VS Buttons
The VS's buttons are used to position the cursor within the name field. Characters can be selected with the →Data Dial.

Mouse
Clicking the desired character places it in the name field and causes the cursor to move one position to the right.

External Keyboard
Working with a →Keyboard PS/2 expedites name entry. The keys' functions correspond to those of the VS's buttons.

Navigate
→Positioning

New Project
→Project New (p. 168)

NEXT
This button moves the →Current Time (→Timeline) position, depending on settings, either to the next Phrase border (→Phrase) or the next Marker (→PREVIOUS/NEXT).

Noise Removal
Noise before and after the desired signal can be suppressed in one of the following ways.

→Phrase Trim (p. 157)
Fade →Phrase Parameter (p. 154)
→Region Erase (p. 190)
Mute via →Automix (p. 19)
→Equalizer (p. 68)
→Noise Suppressor (p. 135)

At the time of printing, the VS didn't provide a denoiser function.

Noise Suppressor
A N. reduces the level of low-level noise (i.e. noise whose level lies below the →Threshold value), making it inaudible. It works like a simple →Expander and is available as a separate block in the following →Effect Patches:

→GuitarMulti (p. 88)
→Guitar Amp Modeling (p. 87),
→Vocal Multi (p. 238)
→Lo-Fi Processor (p. 106)
→DualComp/Lim (p. 59)
→Stereo Multi (p. 211)
→Rotary (p. 197)

Tip
The VS-2400/2000 models allow the use of one →Dynamics processor at a time per channel. By selecting one of the effect Patches mentioned above, and switching off any unnecessary block, an additional, simply equipped Expander becomes available.

Non-Destructive Editing
The VS saves all recorded data as →Takes. These cannot be edited and are always available in the original form*. Audio signals being edited (→Editing) use varying start and end addresses (pointer-based editing). They enable the recall of up to 999 already carried out editing functions (→Undo).

> *An exception to this are re-recorded tracks, which cannot be selected as →V-Tracks. They are lost when →Optimize takes place.*

Normalize
→Phrase Normalize (p. 153)

Now
While importing a track directly from a CD, "Now" can be used to set the start time in real-time (→CD Capture).

Null
As the VS-2000 doesn't have motorized faders, the way of capturing a stored level value after recalling a →Scene can be specified (→Fader Match (VS-2000 only) (p. 78)). The "Null" setting means that the stored value is used until the fader reaches the physical position corresponding to the value. This means that the fader appears to do nothing for a certain time.

Numerical Entry
→NUMERICS (p. 136)

Numerical Editing
→Editing using numeric values is the most precise approach on the VS. Firstly, via Track Menu, select the desired parameter or function. The listed →Edit Points can then be entered in one of the following ways:

Dial Or Mouse
Move the cursor to the desired value (or simply click that value) and rotate the dial, or drag the mouse to change it.

Via NUMERICS
Entering measure and beat numbers can be especially easy in numeric form. Move the cursor to "Measure" within a row, then use ≑ to select the measure number. Finally, enter the number as described for →NUMERICS.

Capturing Time Locations
The →Time Display and the →Timeline's position always indicate the "current time". This value can be copied to the selected field of an editing function. This is easy because the →Transport Buttons, →Locator, →Marker and →PREVIOUS/NEXT perform their normal functions even within the menus, whereas the TIME/VALUE dial's function changes.

NUMERICS
On the VS-24xx models, N. can be used for convenient numeric value entry. Use the cursor buttons to move the red line in the →Time Display to the desired time field (SMPTE, measure or Marker). The example below shows how to move the →Timeline to bar 21.

1. Use ◀▶ to move the cursor to MEAS in the time display.
2. Press NUMERICS.
3. Use the numeric pad to enter "21".
4. Press [YES] to move the Project to bar 21.

For the SMPTE field, numeric entry is performed in "Up" mode (i.e. from subframes/right to hours/left) by default. This direction can be reversed with the "Numerics Type" parameter:

Utility → Global1 → Numerics Type

Press F2 (GetNow) to use the current time. Tracks can be selected by pressing one of the (flashing) buttons on the VS or via the menu selected by pressing F1 (SelTrk) (→Track Select).

Other approaches for editing are →Mouse Editing (p. 132) and →Wheel/Button Editing (p. 243).

O

Octave

Aside from its use as a musical interval, O. is also used for the frequency display where it signifies a duplication of a frequency. In this way, the filter quality (→Q) also can be displayed in Octaves. For Q=1.4, approximately an Octave above and below the assigned frequency will be affected (e.g at a frequency of 1KHz, from approx. 500 Hz to approx. 2 KHz). A →Graphic Equalizer divides the sound spectrum into equally sized bands, which can be individually leveled. The Graphic equalizer of the →Effect Patch P031 has 10 Bands in →Octave widths at its disposal.

Offset

A Timecode offset indicates a time shift between Master and Slave during →Synchronization. It serves to either later align Audio events from Master and Slave or to consciously shift them (e.g music on the VS to an event on a synchronized video tape).

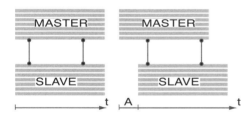

Fig.: Time shift (A) between Master and Slave

It is to differentiate between the production of timecode offsets in Master and in Slave.

Offset (Midi Clock)

Unfortunately, the VS is unable to generate timecode offsets for the outgoing →MTC. It is only possible to generate an offset for the outgoing →MIDI Clock. For this function use the →Sync Offset Time parameter.

> UTILITY → SYNC Prm → Sync Offset Time

Slave Offset (MTC)

The VS can be externally synchronized solely by →MTC. Using the Parameter →Display Offset Time, the incoming Timecode can be delayed in the Slave (VS-2000: Display Param.):

> Utility → Project Parameter*:
> Display Offset Time

Here, to provide clarity, the →SMPTE time can either be displayed according to the incoming Timecode (absolute) or taking the offset time into consideration (relative).

> Utility → Project Parameter*:
> Time Display Format

Offset For Special Applications

Without any further calculations being necessary, different positions in Master and Slave can be easily aligned. This is achieved with an Offset setting of the Slave in relation to the SMPTE display.

Example 1:
The slave needs to start 20 seconds later:
Offset = 20 s

Example 2:
The time position of 40s in the Slave should correspond to 00:00:00:00 in the Master:
Position the Slave to 40s
Change the offset value till 00:00:00:00 is shown in the SMPTE display.

Offset = 23:50:20

Example 3:
A recording begins in the Master at 34:14:0:0 but is only to be recorded after the time position 19:40:00:00 in the Slave.
Position the Slave to 19:40. Via Offset, set the SMPTE display to 34:14:00.

Offset Level

For a mixer channel stereo coupled with →Channel Link or →Fader Link, whose volume is controlled with one single fader, separate volume levels can nonetheless be individually

Offset Pan

adjusted. This is used not only to align "leveled" mixes, but also for sound configuring applications. This can be set via the sub display in the →CH EDIT menu.

VGA:	Click on Sub Disp under Fader
LCD:	position over Fader value + YES

Automix:
In linked channels, Fader settings here produce the →Automix data "OFFSET LEVEL". These will be managed separately from the Volume data. These parameters can be accordingly chosen for graphic editing.

Offset Pan

For a stereo channel coupled via →Channel Link, the individual panorama settings can be separately carried out. This can be set by calling up the sub display in the →CH EDIT menu.

VGA:	Click on Sub Disp under Fader
LCD:	position over Panorama + YES

Automix:
In linked channels, panorama settings here produce the →Automix data "OFFSET PAN". These parameters can be accordingly chosen for graphic editing.

Operating System
→System Software (p. 219)

Operation Display

When working with an additional →VGA Monitor, the display currently being edited is referred to respectively as the O. or the →Operation Target (which is normally the VGA monitor). The →Cursor appears only in this display and the current operations are carried out here. To change between the Operation and Information displays, the following steps need to be carried out:

VGA:	Utility → System Parameter2* → Op.Target
VS-2480:	SHIFT + UTILITY
VS-2400:	SHIFT + MASTER EDIT
VS-2000:	HOME + F6

* VS-2000: System Parameter3

The →Information Display (the LCD) can be selected to show additional information such as →Channel View, →Waveform and →Analyzer.
To do this, the page 3/3 needs to be called up via PAGE in the playlist and the appropriate Display needs to be assigned. These can be secured against a page change in the operation display by using Lock.

Tip:
Firstly, the VGA monitor needs to be activated and the refresh rate set:

Utility → System Parameter2 → VGA Out / Refresh Rate

Operation Target

When in operation with an additional VGA display, the parameter O. determines which of the two displays will be used for editing: either the VGA monitor (which is normally the case) of the internal LCD display (in cases where no VGA monitor is available, e.g during Live Editing in realtime). The chosen display is then named the →Operation Display and operations can be carried out with the mouse. Changing the editing display can be easily carried out directly in the VS or via the menu →System Parameter with the following key combinations:

VS-2480:	SHIFT + UTILITY
VS-2400:	SHIFT + MASTER EDIT
VS-2000:	HOME + F6
Utility Menu:	System Parameter2* → Operation Target

* VS-2000: System Parameter3

Optical
→S/PDIF (p. 199)

Optimize
→Project Optimize (p. 168)

OUT

The →Edit Points IN and OUT determine the beginning and end of an editing region for

→Region Edit, →Automix-Editing and →Region Arrange.

Output Impedance

In order to avoid sound corruption, the input impedance (alternating current resistance) of the devices to be connected (recording gear, amps etc.) should have a higher value than the VS's output. The symmetrical outputs of the VS models possess an impedance of 600 Ω. In particular, this concerns the VS-24xx, whilst the unsymmetrical aux and monitor outs of the VS-2000 (cinch) produce an impedance of 300 Ω.

Output Assign

Internal signals can be sent to external devices via various physical →Outputs of the VS. Because →A/D Conversion isn't carried out until the output section, the routing procedure for analog and digital output sockets are identical. Established routings can be saved, called up and used for other projects (→EZ Routing).
Output routing is carried out in the EZ Routing's "OUTPUT ASSIGN" field.

EZ Routing → Output

The following diagram shows the OUTPUT ASSIGN display:

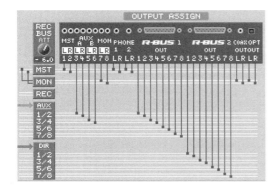

Fig.: Output routing in the VGA-Display (VS-2480)

In the VGA, the assignment of inputs, outputs and effect busses are collectively displayed on one page (EZ Routing → Routing View). The output sockets are represented here within the dark grey field by a photo-like display. The eight plug sockets, phone sockets, both R-Bus connections as well as the digital COAX and OPT outputs can be identified (from left to right). The internal →Busses of the VS are represented by horizontal white lines. The signal flow in the respective busses goes from left to right. Virtual cables, represented here by black vertical lines, record a connection of the output sockets with the respective busses. The signal flow then runs from the busses up into the physical output sockets. All busses, virtual cables and outs are available in a stereophonic form, which doesn't disband in allocation. In the diagram displayed above, the Aux bus pair 1/2 is routed to the outputs AUX 3 and 4 as well as R-Bus1 Out 1/2. In this way, necessary Effect-Sends for →External Effect Processors can be generated.

Direct Outs

Via →Track Direct Out, both analog and digital output sockets, or as the case may be, attached →R-Bus interfaces are equally able to be used for single track outputting.

MonitoR-Bus:

The →Monitor Bus can be applied in various ways and can itself be connected to a further R-Bus pair. Graphically, its exceptional position is presented through the position in "Output Assign". Its access matrix is to be found on the left next to the Bus symbol.

Outputs

The VS's outs are provided in the form of output sockets and serve to output all types of different signals to amps, headphones, →External Effect Processors, monitor equipment, recording devices etc. Not only →Analog Outputs, but also digital outputs are available (→Digital Inputs and Outputs), which, through the relative routings can, in most cases, be freely configured. In the diagram, the connection of devices to the analog output sockets is depicted:

Overdub

1. Powered Monitor speakers
2. External FX Processor (Send)
3. External Master Recorder (Cassette, Dat.)
4. Headphones

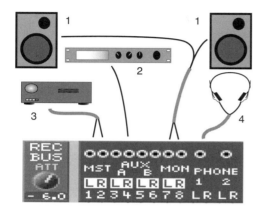

Fig.: Output Setup in the VS-2480 (analog)

Overdub

A new →Recording on free tracks to already existing recordings is called an O. For this procedure, a special →Monitor Mix can be created in order to emphasize single instruments (e.g Bass drum for a bass Overdub). If need be, a test can be carried out beforehand (→Rehearsal).

Overlap

By a multiple →Phrase Copy (Time > 2), if the phrase length is bigger than the distance FROM-TO, the function O. shortens the phrase endings automatically to the required length. The TO times are then exactly positioned (→Quantize) and correspond exactly to a multiple of the FROM/TO distance.

Example:
A phrase, intended for multiple copies, was not trimmed exactly to a bar and is now 1/16 note too long. Already in the 17th repetition, the phrase will be played a whole bar late.

By deactivating the function O., the VS sets the beginning of a phrase directly on the end of the previous phrase. Not trimming the phrase lengths exactly results in a mistake which gets bigger with every subsequent copy made. Only by activating "Overlap" can an exact position of the phrases be guarantied.

Application
Drumloops. In this case, the timing of multiple phrase copies takes priority over the play length.

Tip
Overlap is automatically carried out when Quantize is activated.

Oversampling

The multiplication of the →Sample Rate in the →A/D Conversion to avoid →Aliasing artefacts and to reduce noise is called O. According to the Nyquist theorem, the sample rate must show at least double the value of the highest signal frequency. Because of the conventional low pass filter's slope, higher frequencies cannot be completely eliminated. With the sample rate enhanced, the alias frequencies are projected into the highest frequency range and can then be removed with gentle slope filters. O. is also used during the →D/A Conversion to suppress any disrupting or noisy frequencies.

P

Pad

In →Microphone Preamp, only the VS-2480 provides an additional attenuator of 20 DB via Pad Switch. For all other models, set the Sens knob to the LINE position.

PAGE

In multi paged menus, P. calls up a desired page and assigns the →F Buttons row to this page. Doing this respectively allocates new functions, parameters or pages to the F buttons. In the VGA, the current page and the number of pages can be read in the PAGE area of the footnote (e.g page 2/5 for page 2 of 5). Normally in the LCD, tabs will be used for this function:

Panorama

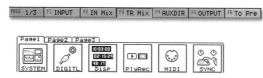

Fig.: choosing a page via PAGE

The appointed F buttons, likewise indicated in the VS →Param1, 2 … can further subdivide pages into sub pages.

Pan Knob Auto Display

By turning the Panorama knob, the respective →CH EDIT menu, when necessary, can be automatically called up. The cursor will then be positioned over the pan field in the page of the relevant mixer channel.

Utility → Global Prm2 → Pan Knob Auto Disp

PAN/AUX SEND Knob

In addition to the regular panorama settings in the VS-2480, 16 encoders can be used for diverse functions. To do this, the following buttons must be activated:

Function	Button to activate	see:
Panorama	none	p. 141
Aux send	KNOB/FADER ASSIGN + 1 - 8	p. 27
EQ and Compressor	AUX SEND/ PRM EDIT	p. 47 p. 68
Midi-Controller	SHIFT + V.Fader	p. 49

For each of these allocated functions, if necessary, the corresponding page of the →CH EDIT menu can be automatically opened (→Pan Knob Auto Display, →Prm Knob Auto Display and →AUX Knob Auto Display).

Panorama

The position of a signal in the stereo image is determined by the panorama parameter. There are different methods available for this in the VS, although display entry is possible in all models. Whilst the VS-2480 has a designated Pan knob, the VS-2400 uses faders specifically assigned to panorama and only the VS-2000 employs a channel strip for this purpose.

Faders for Panorama:

The VS-24xx allows the panorama to be set via the Faders.

VS-2400:

- Activate the PAN button
- Assign the mixer section (→Fader Buttons)

VS-2480:

- Assign the faders for the Parameter entry*
- Button "KnobFader Assign"
- Button 9 (User)
- Assign the mixer section (→Fader Buttons)

    * *System→Utility→Global→Knob/FDR Assign Sw*

Stereo channels (Balance)

Coupled channels are able to be positioned in Stereo, via →Channel Link, in the same way as mono channels. The channel's extreme R/L setting (L63-R63), which, for the time being, is predetermined, is changed solely for one channel whereby the entire image shifts in the corresponding direction. (e.g L10 - R63).
In order to restrict the values (L40 - R40) and for individual allocations (L50 - R33), the parameter →Offset Pan can be pulled up in the sub menu.

Automix

In →Automix, the writing parameters "Pan/Bal" and "Level", are, by default, already activated. For mono channels, the editing of Automix data in the menu "Automix Edit" can be achieved via the parameter "PAN" whereas for stereo channels this must be done via "OFFSET PAN"

VS-2000:

- With CH PARAMETERS (Channel Strip)

Param1, 2 ...

- In the Home Page of the VGA display, the P.knob can be displayed either for the Input Mixer or the Track Mixer.

VS-2400:
In the CH PARAMTER section, the button PAN on the user interface opens the CH edit page of the last accessed mixer channel and positions the cursor over the parameter PAN.

Param1, 2 ...

In many VS menus, not all the prospective parameters are able to be displayed on one page. This is visible by an assignment of the →Function Buttons to Param1, Param2 etc. In this case, for example, the parameter page 1 can be selected with F1, page 2 with F2 etc. The selection can be made either in the VGA display with the F push buttons or directly on the VS with the F button. In this book, the number of the current menu page is shown as follows: "...System 2..." indicates here the parameter page 2 of the system menu.

Parameter Edit

In the VS-2480 the encoder knobs next to panorama entry are also available to be used as a →Channel Strip and for other functions as well*. Operating in this way, the →EQ and →Dynamics parameters of the selected mixer channel can be changed with these knobs. After activating the knobs with the button PRM EDIT, they are assigned to the last accessed mixer channel. This channel number is displayed on the left at the top of the playlist. Additionally, the changed knob function can be read above the level meters in the playlist and also in the CH view of a channel. For a channel from the →Input Mixer or the →Track Mixer, P. can be activated with the →CH EDIT button or in the mixer menu. Knobs 1-5 determine the respective parameters for the dynamics while knobs 6-16 are responsible for the Equalizer.

> *The Knobs can also be applied as →AUX Send controllers for one assigned Aux Bus of 16 channels. This function, however, does not constitute a channel strip.*

For better control when using the knobs, the relevant menu pages can automatically be called up in the LCD (→Prm Knob Auto Display).

Parameter Initialize
→Prm Init (p. 165)

Parameter View

The VS' LCD can also display one parameter for various mixer channels simultaneously, which enables optimal access for certain chosen Parameters. Alternatively, PRM.V. (F6) is able to be activated in the CH EDIT Menu through two methods:

CH EDIT:
→ Select a Parameter → F6 (PRM.V) or
→ F6 (PRM.V) → SHIFT + Cursor ⇕

For further channels, positioning is enabled by cursor ◄► and within the page, by cursor ⇕.
Pressing F6 (CH.V), returns to the normal presentation of →Channel View. The playlist main menu is enabled using HOME.

Application:

- 24 channels: Solo, Mute and Mix-Switch
- 8 channels: V-Tracks, Phase/Attenuator, DIR Paths, Phrase Pad etc.

Tip:
In the LCD display, the eight AUX return channels can be displayed in one page via "PRM.V". This allows a general overview and editing without calling up a separate Return channel.

Parametric EQ

In addition to the channel EQ, the →Effect P. (4-Band →Parametric Equalizer) can be inserted into a mixer channel or the master section, although, and with few exceptions, this will not be necessary. Interesting conclusions though, can come from of studying the preset's →Effect Patches. It is also possible to find interesting approaches to the filtering of different instruments e.g cutting the 300 Hz band for a bassdrum. These settings can then act as guidelines for filter curves, which are personally set with the channel EQ.

Parametric Equalizer	
Effect patch:	P186 - 211
Connection:	2xMono Insert
Application:	Mix

Parametric Equalizer

In a parametric →Equalizer, often both peaking and shelving filters with variable frequencies are available. The VS channel equalizer's two mid bands possess a peaking characteristic (Bell). Desired frequencies can be more advantageously edited here because the area to be raised or lowered is bilaterally restricted by the bell shaped filter curve. Comparatively, the shelving filter can edit only one side instead of both.

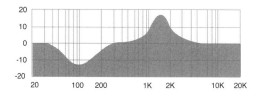

Fig.: Peaking-EQ

The Quality (Q) of a filter determines the strength of its influence on neighboring frequencies during boosting or lowering. The value "1,4" signifies an editing width of approximately one octave above and below the centre frequency (e.g with 1KHz in the area of approx. 500 Hz to approx. 2 kHz). With a lower Quality, more octaves will be affected (a broader frequency band) and with extremely high Q values, an extremely narrow filterband will be controlled e.g only one note. The parameter →Bandwidth displays a more musical description of the Filter's impact using the measurement of →Octaves. In the VS, a filter quality of 0,36 to 16 can be chosen. These correspond to bandwidths of approximately 3 octaves to approx. 1/10 of an octave. As an independent effect, the parametric EQ is also available as the preset P186 - P 211. As well as the actual effect algorithms, most of the →Effect Patches in the VS possess an additional parametric EQ to filter the signals before and after the effect.

* The parameter →Bandwidth(not used in the VS) displays a more musical description of the Filter's impact using the measurement of octaves.)

Partition

To ensure immediate access with the write and read heads to the desired audio data, the →Hard Disk on the VS is subdivided into multiple independent areas, or partitions. These act as independent hard disks and can take up 500MB, 1 GB, 2 GB or 10 GB (recommended). Recordings can only be carried out in the current P. and not spanning multiple partitions. This results in the designation of a maximum recording time respectively for a 10 GB partition. An exception to this is the recording of →Image Files for the creation of an audio CD, which can be carried out spanning multiple partitions.

Deleting Partitions

One single Partition can be deleted in the following way:

SHIFT + Project → Page 4/4 →
Position over IDE:... → F3 ClrPrt

In the →Project List, move the cursor directly over the corresponding partition's name IDE:.... Otherwise hard disk operations cannot be carried out.

Hint:
→Formatting one single partition is not possible. For this the entire hard drive must always be called upon.

Paste

Inserting Audio data from the clipboard can only be carried out in the VS together with →Region Copy and the setting →+Insert.

Patch

Patch
→Effect Patch (p. 65)

Patchbay
In the P., the necessary connections from the input sockets to the input mixer are established via virtual patch cables. The P. is displayed in →EZ Routing together with the physical inputs and contains therewith also the (not visible) →Microphone Preamp including →Phantom Power. Routings can each be carried out in pairs and assigning multiple →Input Mixer channels to one input socket pair is possible.

VGA:	EZ Routing → Routing View
VS:	EZ ROUTING

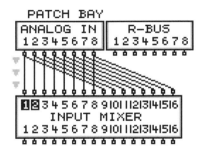

Fig.: The Patchbay of the VS-2400 in LCD

Routing Patch Cables with the Mouse:
- In the top line of the input mixer, click on 1/2 (example)
- Click 1/2 on the patchbay and hold
- Drag to the desired patchbay number.
- Release

Dial/Button Routing of Patch Cables:
- PAGE 1/2 → F2 P.BAY
- Via Cursor ◆▶ choose the Input channel
- With the Dial, establish the desired connection to the patchbay.

Even though, for example, the VS-2000 is operated entirely without a P., it is helpful, particularly in hardwired studio installations. The P. is necessary also for simultaneous recording of a signal onto two tracks. e.g. channels 3/4 of the input mixer are routed to inputs 1/2. The input signal 1 is thereby connected to both the input channels 1 and 3. Channel 3 can now be edited with an insert effect, which will be directly recorded. The still unedited, clean signal is, however, still available on track 1.

Init and Clear (not in the VS-2000)
The connections in the patchbay can be deleted or reset to the default settings.

PAGE 2/2 → F1 IniPB or F2 ClrPB

SELECT (VS-2480 only)
The maximum 16 digital mixer channels stand opposite 20 digital inputs. Therefore a preselection must be made via SELECT. The six SELECT buttons are available each for one input stereo pair:

Button 1 - 4	R-Bus 2
Button 5	S/PDIF coaxial
Button 6	S/PDIF optical

The required stereo input is then to be activated via the buttons, whereupon the buttons light yellow.

Tip
In the INPUT level meters only the levels of inputs actually routed in the patchbay are shown. When dealing with alleged level problems, the correct connection from an input to an input mixer channel must be controlled in this way.

Pattern
1. In →Automix, Patterns serve to save 9 different automix versions within one project.
2. The →Rhythm Track of the VS-2000 uses multiple bar drum patterns to create a rhythm arrangement (song).

PCI Card
→RPC-1 (p. 198)

Peak

A short, often noticeable, level peak in an Audio passage is called a P. With digital audio signals, P.s are responsible for the maximum volume setting as the absolute modulation amplitude boundary of 0 dB cannot be overshot. To increase the →Loudness it is therefore recommended to lower the level peaks via a →Compressor or a →Limiter. P.s cannot be eliminated with →Normalize.

Peak Hold Sw

→Peaks can be frozen in the →Level Meters. When doing this, the current maximum level stays as a horizontal line in the display until it is overwritten by the next peak value. With this function, information regarding the recent peaks of all recording tracks can be obtained.

VS-24xx: Utility → Project → Display → Peak
VS-2000:   Utility → Display → Peak Hold Sw

Peak Indicator

The indicators →Input Clip (p. 96) serve as Ps in the VS.

Peaking

P. is a filter characteristic of an →Equalizer filter, through which a frequency area to be boosted or lowered is restricted bilaterally through a bell shaped filter curve.

Phantom Power

Condenser microphones need an additional terminal voltage input for operation, which is supplied through multiple usage of the symmetrical microphone cable (→Balanced). This feeds not only the capsule with the necessary voltage but also the built in capsule amps in the microphone. Practically two oppositely charged Voltages each of 48 V are fed over resistors into both the audio wires, whereby there is no potential difference between them except for the shield. Because of this, a (symmetrical) connection of dynamic microphones is possible, although defective cables can nonetheless lead to damage.

In the VS, the phantom power supply can be activated in the individual models as follows:

VGA: or:	EZ Routing → R. View → Patchbay click → Phantom. Mixer → CH View → Input INx
VS: or:	EZ Routing → Patchbay Utility System1 → Phantom Sw
VS-2400:	SHIFT + EZ Routing
VS-2000:	Switches on the device's rear panel

In the following diagram, the phantom switch is depicted in an input channel. This channel is routed in the patchbay to the analog input "AIN 1" (Analog Input 1).

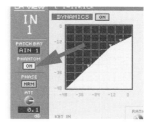

Fig.: Phantom switch in CH Edit (VGA)

Phantom Sw
→Phantom Power

Phase

With Phase switch, the Phase position of a signal can be rotated up to 180°. Particularly when multiple microphones are used, the →Phaser-like →Comb Filter Effect can be suppressed through rotating the phase position. A typical example of this is found in micing a snare with a microphone at the top of the snare drum (membrane) and an additional microphone under the snare, with which an oppositely charged electrical voltage is produced. To avoid frequency dependent deleting, the phase can be respectively inverted in *one* microphone (normal/inverted). In the VS, "Phase" is activated on the →CH EDIT page:

CH Edit → Phase

Phase Cancellation

Tip

The volume of an individual in the mix can be subsequently lowered through the inverted addition of a single track to the generated master tracks (→4. Correcting A Mix Without Redoing It (p. 127)).

Phase Cancellation

Combining two signals with the same frequency but 180° out of phase will result in some frequencies being cut, whilst others may end up being boosted. If, on the other hand, similar frequency spectrums meet each other in delay, certain frequencies will be deleted in regular intervals while others will remain unchanged (→Comb Filter Effect), which results in nasal sound characteristics. This can be corrected by activating the →Phase switch and changing the position of the microphone.

Phaser

The P. is one of the oldest →Effects, which takes advantage of the deletion of phases. It generates moving, smooth sound characteristics with minimal dissonance or going out of tune and is used, therefore, principally for Guitar chords, organ sounds and pianos. The connection takes place as a →Send/Return Effect (p. 204).

Working principle:
The signals are firstly split into two signal paths. After shortly delaying one path, it becomes modulated by an oscillator and the delay time changes. Both paths are finally mixed together, which produces typical phase cancellations (→Comb Filter Effect). The allpass filters, which are used for delaying, have a frequency dependent impact. This causes certain frequencies of the signal to be in or out of phase with each other. The resulting Comb Filter effect, in contrast to the →Flanger, has, therefore, no periodicity. The number of allpass filters can be determined with the parameter "Mode" (Stage). In the VS, the →Effect Patch →St Phaser and →AnalogPhaser are available for the phasing effect.

PHONES

The stereophonic output P. serves the connection of customary headphones to hear user defined bus signals. In the default settings, the phones out is routed to the →Monitor Bus (in the VS-2400 irreversibly), which is in turn routed to the →Mix Bus so that the headphone signal is identical to the one outputted through →MASTER OUT. Because, in this case, a series connection is provided, setting the phones volume is to be carried out as follows:

- Position the →Monitor knob to approx. 0 dB
- Set the desired headphone volume with the PHONES knob.

Controlling Monitors and Phones separately

In compliance with →Output Assign on the OUTPUT page in →EZ Routing, the phones out can also be routed to an optional bus or direct path (not in the VS-2400). In this way, in the VS-2480 and VS-2000, the headphone and monitor volumes can be separately controlled.

Routing the Phones out to the Master-Bus:

- Here, the monitor out is balanced via MONITOR, and the headphones via the PHONES knob.
- The MASTER fader is responsible for the internal (→Master Tracks!) or external recording of a mix.

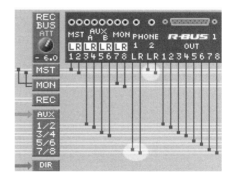

Fig.: Headphones routed to Busses (VS-2480)

This diagram clarifies the independent volume control for the entire bus from Phones 2 (routed to the MasteR-Bus) and the monitor signal.

Phrase Copy

Additionally here, to create an independent →Headphone Mix for inputting musicians, Phones 1 is connected to Aux Bus 7/8.

Phrase

A defined audio passage from one track is called a P. These are represented in the playlist by a continuous bar. The P. is distinctly determined by its length which in turn determines the play time with the beginning and end of the bar. This allows the arrangement to be optically controlled. Once created, phrases can be particularly used effectively for repeated →Phrase Editing.

Phrase Copy

Copying a limited track passage (→Phrase) to a desired time position and an optional V-Track (→Phrase Edit) is called Phrase copy.

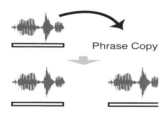

Fig.: copying a phrase

This virtual copy of the passage doesn't require any additional memory space on the hard disk (→Editing). The fact that the passage is already time defined by the beginning and end of the phrase, proves to be advantageous when using phrase copy because the respective bar will be observed in its entirety. To copy, merely specify the point of origin →FROM within the phrase and the target time point TO. Normally when using →Mouse Editing, FROM and TO are assigned graphically by dragging and oftentimes don't have to be watched over by the user. As well as mouse copying, when necessary in the VS →Numerical Editing and →Wheel/Button Editing can also be used.

Application:

- Classic arrangements from segments (e.g. Drumloops)
- Copies of passages for security etc.

1. Graphical copying with the Mouse

This editing form represents a practical method for copying, which can be carried out without any additional scrolling within the playlist. The point of origin and the target time point (TO) should be found within the playlist display. The model procedures for further phrase operations are described here in detail:

• *Activating Phrase Edit*

For a Phrase operation in the VS, firstly activate the mode "Phrase-Edit". Region Edit is already set in the default settings (→Phrase / Region Selection).

VGA-Display

By clicking the red REGION button, it changes to PHRASE (green). The field is to be found under the master level meter on the right hand edge of the playlist.
Or: right click in the playlist "PHRASE"

LCD

By right clicking in the playlist, a flip menu opens. PHRASE can be chosen here.

VS-2480

Only the VS-2480 offers a designated button for toggling between the PHRASE/REGION.

• *Phrase Selection*

One click on the desired phrase selects it and prepares it for further operations. The phrase will be displayed inverted. After releasing, the mouse cursor changes to (Hand, open).

• *Click on Phrase and Hold*

The cursor changes to the icon "Hand closed" and the Phrase is ready to be moved (→Phrase Move).

• *Additionally press SHIFT*

To confirm the activated Copy function, a "+" appears in the mouse cursor.

• *Dragging a Phrase*

Hold down the mouse button + SHIFT and move the phrase copy to the desired time and respective track (use →Quantize for a grid).

Phrase Copy

• *First Mouse button, then let go of SHIFT*
The copy will be carried out.

Application for Mouse Editing:
This form of graphic copying is well known from classic software sequencers. It enables a quick and clear application.

Tip
For fairly remote TO positions and for extremely accurate copying, it is more advantageous to use either →Wheel/Button Editing or →Numerical Editing.

2. Graphical Copying via Time/Value wheel
→Wheel/Button Editing combines graphic clarity and the comfort of using a mouse with the accuracy of numerical editing. In the VS-2480, the Mouse isn't needed at all for this function, whilst in the other models, the mouse is only required for one step of production.

• *Deleting TO from the Edit Storage*
SHIFT + CLEAR + TO
Details are described under →Edit Points (p. 61), →Edit Point Buttons and →CLEAR.

• *Selecting Phrase specially*
Use the cursor button to call up each track, which contains the phrase to be edited. Then position the position line over the desired phrase and click YES (→Phrase Edit (p. 150)). The phrase turns black. In a stereo coupled track, both phrases will be activated.

• *Choosing COPY*
To choose Copy, right click in the playlist and select COPY (VS-2480: COPY button)
A graphically marked copy with the dimensions of the original phrase appears.

• *Positioning the Copy*
Time: The time positioning of a copy can be comfortably carried out with the wheel. In →Time Display the cursor (here a red line) can be simply positioned over the desired time mass (hours to subframes) via ◀▶. Furthermore, Bars, beats and also Markers/Locators and all methods of →Positioning can be used.

Because in this case the graphical copy can't be immediately seen, turn the wheel respectively one grid position forward and one back.

Choosing a Track: The cursor buttons ▲▼ move the graphic copy to the desired track.

• *To Copy*
At the signified position, YES carries out the process of copying. In the LCD, before the operation is performed, a numerical control can be activated, which lets additional copy parameters be set (→Edit Message, via right clicking in the playlist).

Application for Wheel Editing:
When coping to fairly distant target positions, which otherwise would require scrolling through the playlist.

Tip
Quantization of the TO position to whole or, in other cases, quarter notes, can be achieved by through entering them into the respective fields "MEAS" or "BEAT" in the →Time Display.

3. PHRASE COPY with Numerical Entry
Using →Numerical Editing (p. 136) allows extremely accurate copying and provides a concise table for a general overview.

Application:
- For TO positions which are far away.
- For copies based on a score
- For postproduction with predetermined SMPT times.

• *Menu "Phrase Copy"*
Firstly choose "Phrase Copy" in the TRACK menu (LCD: →Track Phrase Edit Menu).

• *Entering FROM and TO*
Enter the →FROM* and TO values (→Edit Points) with the dial or by clicking and dragging with the mouse.

> ** The time position for →FROM doesn't have to correspond to the beginning of the phrase. In principle, the position could assume any value within the whole phrase.*

Phrase Divide

Tip

In the VGA, the Playlist operations remain further active. In this way a combination of numerical and graphical entering of time values is possible:

- →Positioning at the desired destination time for the copy (e.g. via →PREVIOUS/ NEXT or via mouse in the →Time Display). This time position can be assigned for TO via →Edit Point Buttons:
  SHIFT + TO (VS-2480: without SHIFT). The value is then transferred in the TO field and can be corrected here via dial.
- Clicking on a phrase creates FROM at the phrase start and TO at the clicking position. This selects the phrase automatically.
- FROM and TO can also be varied by dragging the respective icons in the →Measure Bar and can be corrected later numerically.

• *Specifying the Source/Target tracks.*
Source Track: The source track must correspond to an active (visible) phrase. Call up the menu "SelPh" with F1 to check this and mark the source track with F3. Clicking on a phrase also selects the source track.
Target Track: By default, the same Track number is selected for source and target. Other tracks can be easily selected with the value dial. By holding SHIFT, the selection is carried out only with a step wide of 16 V-Tracks.
Source and target tracks can also be directly entered with the blinking SELCT and STATUS buttons on the VS:

| SELECT button | → | Source-Tracks |
| STATUS button | → | Target-Tracks |

Using F1 (Back) returns to the main page.

• *Multiple copies*
For multiple copies the following parameters must be set:
Copy Time: Specifies the number of copies for the Time position TO.
Quantize: When making multiple copies of a phrase (Copy Time > 2), the VS corrects the position of the TO point automatically to the nearest quarter (→Quantize). Because of this, the rhythmical structure is kept, which is necessary, particularly when working with a synchronized midi sequencer.
Overlap: With multiple copies, →Overlap guarantees a constant distance between the respective phrases. These distances, however, if necessary, will be shortened and then quantize by the VS.

Using OK or EXEC, the operation will be carried out. Corrections are possible by using →Undo and entering new values. The respective Undo level allows changing back to previous operations. Possible →Error Messages mostly refer to wrongly selected phrases.

Phrase Delete

Deleting an entire phrase can be carried out uncomplicatedly and a time entry via →Edit Points is not required. The position of following phrases doesn't change through this operation. Via →Undo, or in other cases, →Phrase New, the function can be reversed (→Non-Destructive Editing).

1. Position Timeline over Phrase

- Call up the right Track and position the →Timeline over the desired phrase.
- Activate with YES or click on the phrase

2. Deleting a Phrase

- Right click in the playlist and select DELETE (VS-2480: DELETE button).
  Or: TRACK → Phrase Delete → Mark →OK.

Tip
To merely temporarily mute, the Volume levels of a phrase in the VS-24xx can be set to "-∞" (→Phrase Level). To eliminate unwanted passages at the start or end of a phrase, →Trim In/Out can be carried out.

Phrase Divide

Divide, also called Strip Silence, deletes quiet passages from a phrase and generates new phrases for the wanted signals. This results in pauses between the generated phrases, which

149

Phrase Edit

guarantee the time position of the remaining recordings.

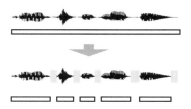

Fig.: Automatic Deletion with Divide.

The diagram above shows not only the wave form but also the bar display in the playlist before and after editing with phrase divide.

Applications:

- Cleaning up tracks (e.g. Toms)
- Noise gate (removing HiHat from a Drum track)
- Regaining memory storage space* etc.

Track → Phrase Divide

The parameter Threshold specifies the boundary between wanted signals to be received and interfering signals to be deleted. Signal rates under this level will be viewed as interfering signals. Passages that are louder than the threshold pass through without being affected. Nonetheless, in order to not cut passages, which are merely being faded, two independent threshold levels are available:

IN Threshold: If the beginning of the sound has a higher volume than this level, it will be interpreted as a wanted signal (→Threshold).

OUT Threshold: After the signal is recognized as a wanted signal through the IN Threshold, the signal level, when fading, can drop out significantly until it finally falls below the OUT Threshold and the quiet noise is interpreted as an interfering signal.

Margin: The generated phrases are extended by the values of IN time and OUT time. This maintains low level regions below the specified thresholds in the beginning and the end of the phrases

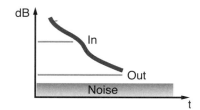

Fig.: Divide In and Out Threshold.

* Through →Project Optimize (p. 168), the memory used for deleted passages and the respective →Partitions of the hard disk become available again. Pay attention to incoming →Fragmentations enhanced through optimize.

Phrase Edit

→Phrases with their predetermined lengths will be favored for repeated operations. This results in permanently fixed start and end times of a passage in contrast to a region, which merely saves these during the current operation.

Fig.: Different Phrase Displays

The editing of a phrase takes place in the same way as the similarly operated →Region Edit by hard disk recorders as →Non-Destructive Editing and can be reversed at any stage with →Undo. The following methods for phrase editing are available: →Numerical Editing, →Mouse Editing and →Wheel/Button Editing. In stereo coupled track channels (→Link), editing of a phrase always takes place as a stereo pair. Make sure that the →Record Modes correspond with each other, particularly in operations involving →Master Tracks (→CDR).

Application:

- Multiple copying (Drum Loops)
- Changing arrangements
- Navigate using the Start and End points of phrases (→PREVIOUS/NEXT).

- Uncomplicated deleting (→Phrase Delete)
- Volume changes and blending of passages (→Phrase Parameter) etc.

Tip
Using the quantizing →Grid for songs inputted via click, simplifies editing considerably. This is due to the fact that here, values can be entered musically in bars and quarter notes and not according to time data.
For single editing, →Region Edit is quicker to use. Multiple copying can be generated via →Region Arrange.

Creating Phrases
Phrases are generated on the one hand involuntarily through recording within a track (also →Punch In/Out) and purposefully through diverse editing operations such as separating (→Phrase Split, →Phrase Divide), creating (→Phrase New) etc. Phrases can also be created by copying track passages (→Region Copy) to any time positions and V tracks. The length of phrases can easily be changed subsequently with →Phrase Trim.

Editing Phrases
Each phrase operation is carried out using the three following steps:
- Activate the Edit mode "PHRASE"
- Select the Phrase
- Carry out the appropriate operation

1. Activate Phrase Edit
Click with the mouse over the REGION button or right click in the playlist 'PHRASE'. In the VGA menu "TRACK", select the desired phrase operation. In the LCD menu "TRACK", change to TRACK PHRASE EDIT with F6 (→Phrase / Region Selection)
VS-2480: PHRASE button

2. Selecting a Phrase
Selection, in the simplest case a mouse click, prepares the chosen phrase for the desired edit operation. The position within the particular track and the play length will be automatically read out, which would require further procedures when using →Region Edit. A phrase that has been selected in this way is shown in the VS display as a bar with dotted boundary lines and in the VGA as a completely black bar with a frame in the same color as the track.

- Click on the desired phrase or:
- 1. With the Cursor ⇕ choose the track
  2. Bring the →Position Line over the Phrase
- Press YES

To select multiple phrases:
Cursor Up/Down + YES
or: Click with the mouse + SHIFT
or: Click on a free Track and drag it vertically over phrases.

3. Carrying out a phrase operation
When a phrase has been selected in this way, the respective operations can be carried out.

Phrase Export
P. Export stands for the conversion of a Roland audio format →Phrase into an all-purpose wave file and burning it to a CD-R. In this way, passages that are recorded in the VS can be further edited or mixed etc. in other audio systems.
- Insert a blank CD-R
- Position →Timeline over the phrase that is to be exported
- Call up the Export menu:
  CD-RW/MASTERING → Phrase Export
- Only one selected phrase is visible and can be markered using F3 (Mark). Further information can be obtained via F4 (→Info)
- Phrases in Stereo Tracks are converted into stereo wave files.
- F5 (Next) calls up the CD menu to select the write speed and the highly recommended inspection (Verify = On).
- OK starts the burning of the Wave files to the CD-R.

Tip:
- Whole →Tracks can be burned onto one CD with the operation →Track Export.

Phrase Level

- Tracks can also be transferred in realtime in a computer with the appropriate audio software (→RPC-1, →V-Fire).

Phrase Level

The volume of single phrases is leveled independently in the menu →Phrase Parameter (p. 154) (not in the VS-2000).

Track → Phrase Parameter → Parameter

Phrase Mode

In the VS, two methods are available for audio editing. Whilst →Region Edit is already activated as a default setting, it is possible to change and toggle to →Phrase Edit and back via the diverse procedures of the →Phrase / Region Selection (p. 157).
Example: With one click in the REGION field (red) on the right in the playlist, the mode changes to PHRASE (green) and vice versa.

Phrase Move

Moving a phrase to a desired time position and any V track (→Phrase Edit) is called P. This operation can be carried out as an analog operation corresponding to →Phrase Copy.

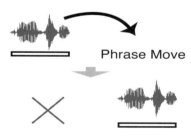

Fig.: moving a phrase

It solely differentiates itself by deleting the originals at the place of origin and by the fact it is missing the multiple copy function (*Times*).

Track → Phrase Move

Application:
- Moving passages
- Correcting rhythmical inaccuracies

Phrase Name

When generated, phrases are automatically numbered hexadecimal. Giving phrases a more significant name can make identifying them easier with extensive and repeated editing work (→Name Entry). Phrase names are displayed in the respective phrase edit menus via F1 (→SelPhr) and additionally in the playlist of the LCD by the chosen track. For entering the names, it is thoroughly recommended to use an optional →Keyboard PS/2.

Track → Phrase Name

Application: For better identification by extensive use of phrase edit operations.
VS-2480: By activating →Phrase Pad, the respective phrase name will be shown in the VGA instead of the track name.

Phrase New

If phrases have been changed or are believed to be lost, the original recordings (→Takes) can usually be retrieved. This is always the case if the operation →Optimize still hasn't been applied to the current project. Phrase New generates a virtual copy of the (unchangeable) take, which can then be used as a new phrase. In accordance with →Non-Destructive Editing, editing operations don't access the original but rather always a copy of the take. In this way, even after a deleting operation, the original recording remains untouched and any number of new phrases can be generated from it at any time. In contrast to →Undo, with which the entire project's state including *all* recordings and editing steps is reset, Phrase New allows a single phrase to be restored in the current stage of the project.

Application:
Reactivating deleted and edited recordings or recordings that are no longer able to be found.

Track → Track *Phrase Edit Menu* → New → *"Phrase NEW"*

In the TAKE field, firstly choose the desired original recording. Not only the length (Start/End), but also the production date can be read out. Using TRACK, the target track for the phrase to be created will be specified. With a step range of 16 V tracks, selection is carried out with SHIFT. The new start time is also able to be freely allocated in the →TO field. To make things simpler when transferring the original take start time, two further functions are available in addition to *Orign (F4):* →Get Now used for entering the actual time position and →Go To for positioning the project over the respective time position. With OK, the new phrase will be produced at the corresponding time position and on the respective V track. Limitations for generating arise in relation to the →CDR of the master track because the →Record Mode of the phrase being created and the target track must correspond.

Take List
Searching for takes proves to be a lot easier with the TAKE LIST, which is to be called up with the function button F1. Displayed in table form, takes can be called up according to date of production (*Hist*), sorted by name and V track, previewed (*PreViw*), and call up production information (*Info*). Whether or not a take is accessible at all is able to be read in a black box on the left next to the name. It is not possible to delete unwanted takes in the take list. Instead for this function, use the →Take Manager, which is to be found in the phrase menu.

Phrase Normalize
The N. function boosts the level of a Phrase up to a maximum value of 0 dB - for the highest →Peak in this Phrase (→SelPhr (p. 204).

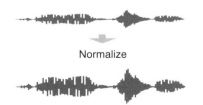

Fig.: Normalize

A physically new phrase will be generated here and used instead of the original. It does not change the quality of the digital signal and can therefore not take advantage of the high →Resolution to be found in higher recording levels.

Applications:
- Level boost of low dedicated phrases
- Enables the use of a compressor for low level signals

N. first looks for the highest level used in a →Phrase (example: –12dB). Next, it raises the level of the entire Phrase in such a way that the peak value equals 0dB (example: +12dB). The VS performs this operation on a copy. The original is still available as a →Take and can therefore be recalled using →Undo. Move the →Timeline to the desired phrase and select the "Normalize" menu:

Track → Phrase Normalize

Only the Phrases below the →Timeline are shown in the list.
- Select the Phrase to normalize by clicking it or pressing F3 (Mark). Use F2 to select/deselect all Phrases below the timeline.
- Use OK and answer the confirmation message to start the operation. Depending on the Phrase length, this operation may take some time.

Tip
The phrase parameter's →Phrase Level can also be used to change a Phrase's level (→Phrase Parameter).

Phrase Pad
Only the VS-2480 provides an integrated sampler with its "Phrase Pad", which bypasses audio tracks instead of the recorders either manually or with the →Phrase Sequencer. The sampler function of a track can also be used together with the normally played track. A mixed operation is also possible. The first phrase of the track will always be used as the sample.

Phrase Parameter

Application
Immediate playback of audio material additional to the playing tracks for theater, circus presentations, flourishes etc.

The following procedures enable playing tracks manually using the phrase pads:
1. Call up →CH EDIT for the desired track and then activate "Phrase Pad".
   - *Gate:* Plays only, if the button is held
   - *Trigger:* Press once for start and stop
   - *One Shot:* After pressing it stops at the phrase end. A restart is possible.
2. Activate the PHRASE button
   To do this, "Phrase Pad" must be activated in at least one track.

Playing a Sample
To play, press the corresponding button "TRACK STATUS / PHRASE PAD". For visualization purposes, the phrase name appears in the playlist of the VGA instead of the track name. The phrase is displayed colored orange.

Tip
For an immediate start, a phrase needs to be cleanly cut at the beginning of the music (→Phrase Trim).

Phrase Parameter
With the exception of the VS-2000, in addition to phrase volume, other settings for fading in and out can also be carried out in this menu. Because of its function as a parameter purely for playback, audio data won't be changed and the original phrase remains untouched. After activating the mode "Phrase Edit" (→Phrase / Region Selection), a phrase firstly needs to be selected (p.158). The page "PHRASE PARAMETER" will then be called up using F1 *(Param)*. In doing this, the name of the phrase to be edited is shown.

Track → Phrase Parameter → Parameter

1. PHRASE LEVEL:
The play level of a decided phrase can be balanced with a range of -∞ to +6 dB. In contrast to →Phrase Normalize and depending on the original volume, boosting can lead to signal distortions.
Application: Change volumes, to save manual or automix control procedures in the mix.

Tip
- When using →Attenuator (6 dB), Channel fader (6 dB) and PHRASE LEVEL (6 dB) together, level boosting of 18 dB for recorded signals is possible.
- Dividing a passage into multiple phrases enables leveling each single phrase (→Phrase Split). (e.g unbalanced vocal passages, generating words as phrases and individually levelling them).

2. FADE:
For fading in and out in phrases, FADE IN TIME and FADE OUT TIME are available. The fade IN starts at the beginning of the phrase, the phrase OUT ends at the phrase end*. Fading times can be set between a range of 1ms up to 1 min. (although to avoid clicking noises, less than 2 ms isn't recommended.) Using FADE CURVE, either linear (less expressive) or exponential (aurally correct) curves are able to be created.

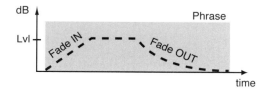

Fig.: Phrase fade in and outs

In the diagram, Fade IN is displayed with a linear envelope and fade OUT with an exponential one. "Lvl" corresponds here with the phrase parameter PHRASE LEVEL in dB.

* *In case the entire time between the chosen IN and OUT time overshoots the actual phrase play time, the IN time will be automatically reduced.*

Phrase Sequence

Application:
- Fading in and out of passages
- Eliminating →Pop Noises etc.

Tip
Because fading within a phrase can't be carried out, the phrase can then be divided (→Phrase Split) at the respective time position. FADE IN can then be generated for the starting point of the new phrase and FADE OUT for the end.

Phrase Sequence

In addition to playing back VS-2480's phrases via →Phrase Pads, the P. allows to record phrase events. This recording can take place in realtime as well as with "Step by Step". Similarly to a midi recording, the actual samples themselves aren't recorded. Instead, Phrase Sequence solely uses commands (→Events) for registering the pressing and releasing of buttons (without velocity value).

Arming Tracks
After turning on the "Phrase Pad", the sequencer is to be activated for the desired tracks. The PHRASE button lights red. With "Phrase Sequence" activated, the colors of the additionally lit up CH EDIT button gives information about the status of the sequencer. This can be changed by pressing the button or clicking in the respective "P" field in the →Playlist, in the →CH EDIT or in the menus →TR F/P or →MltChV:

    Orange: Manual
    Red: Write (record)
    Green: Read

Comparable with the recording of audio signals, to record sequence events, not only the corresponding tracks need to be armed but also the comprehensive sequence records need to be activated.

1. Position PHRASE SEQ in the desired Tracks to WRITE

2. Activate the receptivity for:

    VGA: click in the P ■ field
    VS-2480:   press PHRASE again

The button Phrase blinks red and "PHRASE SEQ REC" is shown in the display.

Recording in realtime
- Press PLAY
- Press the corresponding track status button to release the sample.

Recording to any Position:
- To position Events anywhere without rhythmical quantization, deactivate the parameter Auto Locate:

    PAGE 2/2 → F1(AutLoc) = Off

- Move the →Position Line to the desired time or bar.
- By pressing and holding a track status button, this information will be recorded. When navigating to another time, the corresponding bar will be displayed. This method is perfectly suited to position noises if the VS-2480 is used in synchronization with a video recorder in post production (→Video Recorder Synchronization) or for other such applications.

Recording Step by Step
This method allows events to be entered when the sequencer is stopped. After each entry, the timeline is positioned one defined step further. A step can either denote whole bars (Measure) or also →Beats (normally quarter notes). Pauses (Rests) of the same step width can also be used to position forwards or via Backstep (BckStp), backwards. The setting "Length" is an exception to this. Here, after releasing, the timeline is automatically positioned at the release event, otherwise or in other words, the end of the phrase.

- Activate the parameter "AutLoc"

    PAGE 2/2 → F1(AutLoc) = On
    Press F2 repeatedly → Beat

Phrase Split

- The step width "Beat" (normally quarter notes) causes the positioning after the entry to be moved one beat position further, while "Measure" positions the position line by bars. To enter a pause, "Rest" is used and for stepping back, "BckStp". The length of notes can also be changed: The parameter "Tie" extends the duration respectively by one step.

Editing

After deactivating "Phrase Seq Rec", extensive editing operations can be carried out. These are principally comparable to →Region Edit and →Phrase Edit and are analog operations. Furthermore, a special →Undo is available here.

1. Undo:
   The Phrase-Seq-Undo is not identical to the 999-Level-Undo for audio editing. It merely allows to toggle between the last and current operations.
2. Quantize
   With quantize, Events can be driven into a rhythmical grid and audio samples can be quantized.

   Activate Region → draw a frame →
   Page 1/2 → F1 (Qtize) →
   with the Grid grid choose →YES

3. Deleting Events

   Activate Region →draw a frame →
   Erase or Cut

4. Copying

   Activate Region →draw a frame →
   click +SHIFT → drag→
   mouse click, then release SHIFT

5. Phrase Trim

   Activate Phrase →
   Drag Phrase End with the mouse

6. Micro Edit
   The listed Events are to be positioned using the parameter *Start* and the parameter *Duration* changes the note length.

Without leaving this menu, tracks can be chosen via *Tr Inc/Dec*.

Phrase Split

Using Split, a phrase is subdivided into two new phrases. The division point is normally inaudible. Merely specify TO for the parameter to be set.

Application:
- Preparation for multiple editing
- Create musically meaningful units*
- Particularly multiple Phrase Copying

  * *As well as Split, →Trim In/Out, →Region Edit and →Region Copy are also used when creating phrases of a desired length.*

1. Activate Phrase-Edit
- VGA: Click the REGION button or:
- Right click in the playlist → PHRASE
- VS-2480: PHRASE button
  →Phrase / Region Selection (p. 157)

2. Selecting the Track and Phrase
- Click on the desired phrase or:
- Cursor button ⇕, position over the phrase, YES (→2. Selecting a Phrase (p. 151))

3. Positioning at a Split Time Location
Dial, Transport button, Marker/Locator →Positioning

4. Entering a TO time
Click on the TO Icon in the →Measure Bar or:
Right click in the bar line → TO'   or
SHIFT + TO button   or:
VS-2480: TO button

Calling up "Split"
- Right click in the playlist → SPLIT or
- Phrase Menu: SPLIT or
- VS-2480: SPLIT button

5. Operating Split
With YES the phrase will be split into two new, independent phrases at the specified time position.

Menu "Phrase Split":

Track ➔ *PHRASE EDIT MENU* ➔ *Phrase Split*

Tip
To generate a new phrase firstly create the new phrase's *End* with "Split". By doing this, the phrase remains selected and creating a new split for the phrase start requires one less procedure. Example: Lowering breathing noises or level adjustments of passages with →Phrase Level.

Phrase Trim

Using TRIM, a phrase can be not only shortened but also extended to its original play time - if this has already undergone a →Phrase Edit. Trimming is carried out separately for the phrase start via TRIM IN and for the phrase end via TRIM OUT. TRIM is easily accomplished either graphically with the mouse or exactly with the button/dial and numerical entry.

Application:
- To shorten (or extend) phrases to the desired length e.g for multiple copying purposes (Riff, Verse, Refrain etc.).
- To cleanly erase traces (breathing noises, hum from an amp etc.)
- To shorten for an overall arrangement (Verse, Refrain etc.).

Activate Phrase Edit
- VGA: Click the REGION button or
- Right click in the playlist ➔ PHRASE or:
- VS-2480:PHRASE button

Graphical Trimming
1. Position the Cursor at the End or the start of the phrase – The Cursor changes to the Trim Icon ⇔(OUT) bzw. ⇔(IN)
2. Click and drag in the desired direction – The phrase is shortened.

Trimming with the Mouse and Dial
1. Position at the new phrase end or start. Then with the cursor, choose the track which contains this phrase. YES activates the phrase.

2. To save the TO time
   Press SHIFT + TO *   or:
   Click on TO in the →Measure Bar   or:
   Right click in the Bar line ➔ TO

   * *VS-2480: TO*

3. Operating Trim:
   Right click in the playlist ➔ Trim In / Out
   Or: Choose Trim or Trim Out in the TRACK menu and confirm with OK.
   VS-2480: TRIM OUT / TRIM IN button

Trim numerically as a Table
1. Call up the menu TRIM IN or TRIM OUT:
   Track ➔ Phrase Edit
2. Enter the time positions for the new phrase boundaries either with the dial or via →Get Now (copy the current time).
3. Select the Track of the particular phrase: F1 →SelPhr, then F3 MARK or TRACK STATUS button (→Track Select).
4. Operate TRIM, press F5 OK to do this.

In →Waveform, the operation will be preferably carried out as follows:
1. Within the phrase, click on the desired trim position. Through this, the phrase is activated and at the same time, the TO point is appointed to the click position.
2. Right click in the playlist.
   Choose Trim IN or Trim OUT
3. Press YES to carry out the procedure.

Phrase / Region Selection

In the VS, two editing methods are available for audio data: →Phrase Edit and →Region Edit. With each editing procedure and therefore the choice between these two methods, a basic choice about the way and the possibilities for editing needs to be made. The choice of the entry method can be made as follows:

Track Menu
With the track menu in the LCD, using F6 can change the operation mode. The indicator changes in correspondence to Reg➔Ph or Ph➔Reg.

Pink Noise

VGA-Display

The mode of operation changes automatically through choosing the respective operation in the "TRACK" menu.
With one click in the REGION field (red) on the right in the playlist, the mode changes to PHRASE (green) and vice versa.
Right clicking in the playlist opens a flip menu where phrase or region is able to be chosen (Also in the LCD):

> Right click in the playlist
> Choose 'REGION or PHRASE

VS-2480: PHRASE/REGION button:
Phrase (green), Region (red)

Pink Noise

The sum of statistically distributed sine waves in all audible frequency ranges is called noise. The spectral intensity thickness that is indirectly proportional to the frequency, decreases in P. In comparison to →White Noise, it has a low emphasis.
The →Generator/Oscillator of the VS-24xx is able to produce both white and pink noise.

Pitch

The frequency of a monophonic signal can be tuned with the →Tuner of the VS-2000. The plug-in →Auto Tune allows the automatic correction of intonation in realtime.

Pitch Shifter

A P. (Harmonizer, Transposer, →Effect) generates one or more additional voices from one monaural voice, which are mixed into the main voice in freely chosen intervals. Further effects are generally inserted resulting in complex sound characteristics. A required interval can be entered continuously (Fine) and in half tone steps (Chroma). In the following →Effect Patch, P. algorithms are available:

→Vocal Multi, →Vocal Channel Strip (Plug-In), →StPS-Delay, →Voice Transformer (separately controllable Formats.)

Pitch Shifter

Effect Patch:	P103 - 104
Connection:	Insert
Application:	experimental

Play

1. Play

Play is generally started by activating the →Transport Buttons' PLAY on the user interface or by clicking on ▶ in the VGA.
Through this the →Phrases positioned via the →Playlist will bypass the →Position Line. In case the recording is carried out with the help of the internal →Metronome, the display in the →Measure Bar will correspond to the rhythmical structure of the song. Play can also be activated by:

- The PLAY transport button on the VS or in the VGA
- →Foot Switch (p. 79)
- →MIDI-start command (p. 122) of a synchronized additional VS, →Sequencers or →Video Recorder Synchronization

Additionally, it is possible to move within the project by using the operations described in →Positioning. Playing or recording with a changed speed can be realized using →Vary Pitch. A →Loop operation within personally defined boundaries is also possible.

2. Track Status

The →Track Status "Play" describes the operational state of "Play back" (REC, Off, PLAY.)

Play Status

The operation "Play" is also described as P. (→Track Status).

Play/Rec Parameter

Generally, Parameters appropriate for recording and play are able to be called up in this sub menu of the →UTILTY Menu.

Playlist

The playlist in the VS represents an integral part of the →HOME page. In the VGA display, it contains the active →Tracks and →Phrases, the →Waveform, the →Position Line, the →Measure Bar, as well as the tools for →Zooming and →Scrolling. The playlist can be called up in every menu with the Home button.

Zoom
In the LCD and VGA, increasing or decreasing the track display is done as follows:

SHIFT + Cursor ⇕ / ⬌

LCD
Due to the limits of the LCD, not all functions are available. The display is changed in the following way:

SHIFT + HOME

By performing the operation stated above multiple times, the following displays can be accessed:

- Normal view
- Suppressed V-Track display
- Big playlist display

Plug-in
P.s represent additional programs for extending the functional range. They can be integrated in existing host programs via a software interface. In the VS to instal effect plug-ins, the optional →Effect Board "VS8F-3" is required, which unlike the VS8F-2 doesn't provide permanently installed effects. For operation, load the desired effect plug-ins from the internal hard disk to the Ram memory of the VS8F-3. Depending on the →Sample Rate of the current project (→New Project) and the computing power required for the respective algorithms, a maximum of two stereo effects can be used simultaneously in a VS8F-3. Because of the open platform, in addition to the work's own plug-ins third party plug-ins are also available. An example of these are: →Auto Tune, →Massenburg EQ and →TCR 3000 Reverb. Once installed, effect plug ins can be connected as →Insert or →Send/Return Effects. Not only in the VGA (fig.) but also in the LCD, the display of the controls is realistically photo-like.

Fig.: VGA Display of a Plug-In

1. Preparations

1.1. System Software
The following list of →System Software is a prerequisite for the use of plug-ins. This software is able to be installed using the SETUP DISK provided with the VS8F-3. Nonetheless, it is recommended to constantly update these by going to the website www.rolandus.com.

Model	VS-2480	VS-2400	VS-2000
Software	> 2.504	> 1.504	> 1.504

After the system update, it is recommended to →Format the entire hard disk with the associated →Surface Scan.

1.2. Authorization
The plug-ins are dispatched on a CD-R. During installation, in addition to copying the effect plug-ins to the hard disk, data for authorization is simultaneously written to the CD-R. (e.g For the effect board "xxx" the plug-in "yyy" has been installed.

Hint:
After installation as an "Authorized Version" and also after correctly un-installing, the plug-in can be used exclusively on this effect board. Alienating this plug-in can then only be achieved together with this effect board.
Whilst uninstalling, the VS deletes the plug-in from the hard disk and writes on the installation

Plug-in

CD-R, *"Installation on this Board possible again"*. Through this form of authorization, the manufacturers prevent on the one hand multiple uses of their plug-ins on different boards and on the other, the installation on this board and a new installation on another VS.

> **Formatting the Hard disk**
>
> Before →Formatting the hard disk, the corresponding plug-ins must be un-installed. Otherwise the acquired plug-ins are irretrievably lost.

2. Installation of Plug-ins on the VS-HD

The respective Plug-in CD must be found on the drive. Either a demo version (Trial) or a complete version (Authorized) can be installed. Firstly call up the page "PLUGIN Install", which shows information about the respective effect plug-ins and the installation partition. In the target Partition, which is displayed under DESTINATION, at least twice the memory space needed for the effect programs to be installed must be available (SIZE). The page AUTHORIZE appears with F5 and YES. In this page, an authorized version can be installed by pressing YES and a trial version by pressing NO. After further confirmation via YES, the VS installs the plug-in.

CD-RW → Plug-in Inst → F5 Instal → Yes

Authorized Version:	YES	→	YES
Trial Version:	NO	→	YES

2.1. Authorized Version

When installing the completely functional and authorized plug-in version, the corresponding authorization data will be simultaneously written to the hard disk and the CD-R (see "1.2. Authorization").

2.2. Trial-Version

To test Plug-ins before purchasing them, special trial versions can be installed. These work with the following limitations: All outs of the VS are muted in irregular intervals and play is stopped while the message, *"You are using a trial version"* is shown. Effect parameters cannot be saved either in the project or in USER Patches. For normal operation, this plug-in can be unselected again via "Unselect". To activate an authorized version, the trail version must be un-installed (see "2.4. Uninstalling").

Producing a Trial Version CD:
The operation of a previously installed plug-in on an "unauthorized VS" is only possible as a demo version. It is however possible to burn backup CDs of an authorized plug-in, which can be given to other users for test purposes.

2.3. Backup

For security reasons, producing a backup CD of the installations CD-R is necessary. This is assigned the same value as the original. It additionally contains one's own produced user effects (see below). Because of the propriety of the Roland data formats, copying to a PC or a Mac cannot be carried out. With a new CD-R in the drive, the page "Plug-In Backup" can be called up using CD-RW → PLUG-IN. This displays a list of plug-ins already installed. F4 followed by security prompts that are to be confirmed copies both plug-in programs and authorization data to a CD-R.

2.4. Uninstalling

When formatting the hard disk or selling plug-ins (only possible together with the Effect-Board), the plug-ins *must* be firstly uninstalled. When using this procedure, the VS changes the status of the installation CD-R correspondingly. Subsequently a new installation in connection with the appropriate VS8F-3 is possible.

Hint

Before →Formatting the hard disk, plug-ins must be uninstalled. Otherwise they will be irretrievably lost.

With the installations CD-R or the backup CD-R in the drive, the page "Plug-In Uninstall" can be called up via CD-RW → PLUG-IN UNINS. Installed plug-ins are shown in a list. Using F5

Plug-in

and after confirming numerous security prompts, the effect programs can be deleted from the hard disk. The VS automatically writes the status changes to the SETUP DISK. The plug-in together with the VS8F-3 can then be installed in another VS.

3. Accessing Plug-Ins

In contrast to the internal effects that are immediately ready for operation, a plug-in is loaded from the hard disk to the ram memory of the VS8F-3.

- In the page effect VIEW[*1], an effect processor labeled with [NO PLUG-IN] can be accessed with its associated F button or by double clicking. The INFORMATION page shows: «NO PLUG-IN Selected».
- With F1 (Plug-in), the list of the plug-ins installed in the hard disk appears. Choose the desired effect program and use F5 (SELECT) to select it.
- F3 (Patch) displays the patch list of the algorithm and, via F5, allows the selected patches to be loaded to the processor. The effect plug in is then ready for immediate use as a Loop effect for the appropriate Aux or FX bus[*3].
- F4 accesses the photo-like editing page.

Limitations arise with the use of the respective second board processor. Most Plug In Effects require the computing power of both processors and can only be loaded in the first processor. The list then shows these patches as grey.

*[*1] SHIFT + F3 (VS-2480: EFFECT button)*

*[*2] When multiple plug-ins are installed, the desired group is firstly preselected using F1 in the plug-in List.*

*[*3] In the default settings, the FX number corresponds to the Aux or Fx bus of the same number (e.g Aux3 → FX3)*

4. Editing Effect Plug-Ins

With the exception of the photo-like parameter display, there is no difference between editing plug-ins and editing conventional effects. Starting on the effects page, select the particular effect and call up the edit page using F4. In →CH EDIT the basic connection is carried out as per normal as an →Insert or →Loop Effect procedure.

5. Saving and Loading User-Patches

Edited effect plug-ins can be saved to 100 user patches in the same way as conventional effects. While the internal user effects are stored in the flash ROM of the effect board VS8F-2, edited plug-ins are saved directly to the hard disk. By backing up the installed plug-ins (see "2.3. Backup"), the corresponding User Patches will also be saved to the CD-R.

Saving:

In the Edit or plug-in information page, SAVE calls up the menu PATCH SAVE. After choosing one of the User storage places with a grey background, the edited effect patch will firstly be saved using F5. Only after this can →Name Entry be achieved using F11.

Loading User Patches:

In the plug-in information page, firstly access the Patch select page using F3 and the USER page with F2. After selection, the desired user patch can be loaded to the processor using F5 (SELECT).

6. Updates

Extending the function's perimeter, improvements and bug fixes can be updated, like any other software, by downloading the updates and plug-ins are no exception. The procedure for downloading the updates and burning an update CD is done in accordance with →System Update. The installation of updates is carried out as an analog procedure. The original authorization CD can only be used subsequently.

Tip

By installing a VS8F-3 in the VS-24xx's Slot A, the assignments of the →Send/Return Effects (Aux1 and 2) for personal and demo songs to Effects 1 and 2 no longer correspond (Generally Hall and Delay). This is the result of the new allotment of the plug-in processors in Slot A. The following procedures allow the created Aux 1 and 2 Send settings to be used unchanged for feeding effects 3 and 4 (the new VS8F-2 processors slot):

- Via →EZ Routing, change the Aux bypass in the respective effect processors (FX3 to Aux1, FX2 to Aux4).
- The new effect patches must be selected manually.

PLY

 In the →Playlist, P. signifies the "Play" status of recorded tracks. Using the associated arrow in the footer of the PLY column, all tracks can be moved to play status.

Pointer Speed

The speed of the mouse cursor in the display when moving the mouse can be set with a range of 1-5 (→Mouse Editing):

Utility → System3 → Pointer Speed

Pop Noise

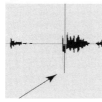 Low frequency, explosive noises from humans, which are created through an air vortex directly in front of the microphone diapraghm are called Pop noises. This particularly concerns the consonants "P" (projected up), "T" and "K" (projected down).

To avoid this:
- Use a pop filter
- Change the angle of the microphone and the distance:

Removal:
- Delete in the menu page →WAVE DISP with (→Region Erase, →Phrase Trim).
- Split the Phrase directly before the noise occurs (→Phrase Split)
  Create a Fade In for the new Phrase (→Phrase Parameter).

Position Line

The current time is displayed numerically in the →Time Display and graphically in the →Playlist through a perpendicular red line, which is called the position line or "Time line". At its position, →Locator, →Marker, →Edit Points, →Automix-Snapshots etc. can be created. The position line itself can be moved using diverse operations (→Positioning). For numerous →Phrase Edit operations, the phrase to be edited is firstly set over the P.

Positioning

Positioning, otherwise known as navigation within a project, can be easily carried out in the VS in diverse ways with the appropriate tools. It enables navigation to a passage for recording, play back, →Editing of Audio and →Automix-data, →Mastering, registering →Locators and →Markers etc.:

- →Transport Buttons (p. 230)
- →Time Display (p. 224)
- →Locator (p. 106)
- →Marker (p. 109)
- →PREVIOUS/NEXT (p. 164)
- Via SHIFT and →Edit Points (p. 61)
- Externally controlled as Slave in a →Synchronization (p. 217).

Pre/Post

P. signifies sending signals to another destination either before the fader (pre) or after the fader (post). This can concern either a →Level Meter, an →AUX Bus or a →Direct Path. Signals are generally sent to a destination via a post-fader-connection, which guarantees a maintained relationship between the original and the second signal's levels, even with different fader settings. Post connection can be applied to the typical →Send/Return Effects such as Hall, Delay, Chorus etc. or also to the amplitude with consideration for the fader position. By contrast, the pre-connection for applications is used if the fader position shouldn't have any influence on the signal being sent. This is important, for example, for the →Monitor Mix. The monitor signal is only determined by the setting of a →Send knob here independently of the sound level in the mix.

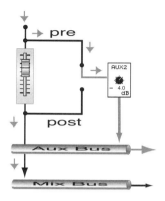

Fig.: Pre/Post connection for an Aux Bus

Master Section:	F1 View
VGA's Channel:	Click on Post or Pre
Duct in the LCD:	Position at Post → YES

Preamp
→Microphone Preamp (p. 122)

Preamp Modeling
The →Plug-in P. is delivered with the effect board →VS8F-3. Both of the independent →Channel Strips provide an algorithm, which simulates the high quality →Microphone Preamps such as Focusrite, Neve, Millenia, Manley, Avalon, Crane Song and Summit Audio. Its definitive sound characteristics such as frequency response and the relationship of odd and even numbered partials and distortion can be defined in the corresponding parameters. With one processor (one board provides two processors), a stereo or two mono effects can be activated. P operates with a →Sample Rate of up to 96 KHz

Pre-Amp Modeling

Effect Patch:	Plug-In
Connection:	2x Mono Insert
Application:	Special

Firstly assign "L" or "R" to the corresponding inserted channel with TARGET (→Insert-Mode). Pre Amp modeling contains numerous effect blocks: The →Compressor reproduces the regular performance of transistor and tube compressors. It also provides a →Soft Knee characteristic. With the →Enhancer/ →De-Esser which can alternatively be used, high frequency rates can be emphasized or lowered. A →Parametric Equalizer enables the corresponding frequency corrections.

Pre-Amp
The desired pre amp is chosen using TYPE and the over tone structure of the sound image is determined with the HARMONIC parameters. Signal rates with a level above the H. threshold generate harmonics whose strengths are controlled with H.Level and the spectral configuration with H.Color. Via WARMTH and BRIGHT additional changes in the low and high ranges can be made. These parameters are already assigned to the simulated pre amps with the choice of a pre amp patch.

Preset
P.s represent procedure settings for →Effects, →EZ Routings and Patterns (→Rhythm Track), which can't be overwritten and which are identified with a corresponding "P". For modification, a Preset must be modified and then saved under a →Name Entry in a free user memory storage space.

Preview
The function P. is used for the exact, acoustic location of a passage's beginning or end that is being searched for. With the help of the found time positions, editing operations can be carried out (→Editing) or →Locators and →Markers can be positioned.

VGA:	Right click in the playlist or Preview
VS:	PREVIEW button

Preview Length

The pre and to time lengths can be set using →Preview Length.

Preview From

"P. From" starts play from the current time position and stops it again automatically after approx. one second (by default). The →Position Line is then set back exactly at the starting point. This allows repetitively starting from a defined time point without having to manually position or use →Locator to start.

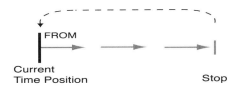

Fig.: Preview FROM

To find a special passage, the dial is firstly used to go forward or back a few frames and then P.From is started as a control until, step by step, the exact position has been reached.

Preview To

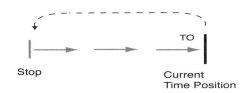

Fig.: Preview TO

"P. To" starts play approx. 1 second before the current time position and stops the timeline then exactly at this time. Doing so allows the end of a passage to be checked.

Preview Thru

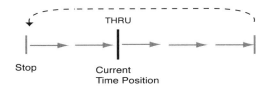

Fig.: Preview THRU

This option is suitable for roughly searching for a time point during playback of the entire passage.

Preview Length

The →Preview listening times can be individually chosen with a range from 1-10 seconds.

Utility → Play/Rec Parameter → PreviewXX Length

Particularly as yet unheard audio material can be found easily with larger length values in conjunction with "Preview Thru".

PREVIOUS/NEXT

This button allows the →Timeline (current time position) to be positioned either at the previous/next marker or the previous/next →Phrase boundary. This setting is carried out using the →Previous/Next Switch. The modes of operation can be activated more simply, however, by simultaneously holding SHIFT.

VGA:
VS: PREVIOUS or NEXT button

1. Positioning at Phrase Boundaries

With this mode of operation, the timeline is positioned within the selected track at the respective phrase boundaries. With Next at the start of the first phrase, then at its end, at the start of the following phrase and then at its end etc., as an analog function and using Previous, it is possible to position at the previous phrase starts and ends.

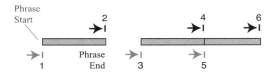

Fig.: Positioning Phrases with Previous/Next

Application

- Exactly searching for a phrase end or start for editing.
- Positioning within the project also for passages which are far away from each other.

2. Positioning at the Marker

The mode of operation called "Marker" allows positioning at the previous or following →Markers.

Previous/Next Switch

Using this setting parameter, the function of the →PREVIOUS/NEXT button changes from the Phrase boundary position to the Marker position.

> Utility → Global1 → Previous/Next Sw

Tip
The other respective modes of operation for the Previous/Next button can be directly activated by simultaneously holding down SHIFT.

Printing Effects
→Effects, recording (p. 67)

PRM EDIT

In the VS-2480 for editing the →Dynamics and →EQ parameters, the control "PAN/AUX" is additionally available. After activating the function using the button PRM EDIT its LED lights red and allows an easy value entry using knobs. The function →Prm Knob Auto Display is used to automatically call up the respective →CH EDIT page. In the VS-24xx chosen parameters can also be controlled using the motor faders (→Knob/Fader Assign).

Prm Init

When an unexpected operation condition appears, the original default settings can be reset by initializing the mixer and/or the utility parameter (→UTILTY Menu):

> Utility → Prm Init

Mixer
By choosing "Mixer", in addition to the channel parameters also the →Routings of the in and outputs are initialized. Scenes, Locator, Marker and the Tempo-Map will not be initialized with this choice.

Utility
Synchronization, Midi and Play/Record Parameters are reset.

Tip
Generating a →Scene before initializing allows a subsequent analysis of problems or mistakes.

Prm Knob Auto Display

In the VS-2480, →Dynamics and →EQ can also be operated via the →Channel Strip (→PRM EDIT). When necessary, turning a "Panorama" knob will automatically call up the respective page of the →CH EDIT menu:

> UTILITY → GLOBAL Prm2 →
> Prm Knob Auto Disp = CH VIEW

PRM.V

In the CH EDIT of the LCD chosen parameters can be displayed for multiple mixer channels (→Parameter View).

Program Change

The →MIDI command C. allows programs in sound modules, effect devices and midi mixers to be changed by an external midi sender. In the VS not only preset and user effect patches (→MIDI Effect Patch Select), but also scenes can be changed (→MIDI Scene Change).

Project

A project contains all the data and settings of a "Song". It can contain one track, a song, all the songs of an entire album or even the songs of numerous albums. In addition to audio data, all the settings for the →Mixer, →Routing, →Effect, →Hard Disk Recorder, →Automix, →System Parameter, and →UTILTY Menu Parameters are stored. A project is primarily limited by the audio data's required amount of disk space. The available number of →Events, however, is also limited. In total, up to 200 projects can be stored in a partition. The following project operations can be carried out from the →Project List although generally these need to be →Marked for this purpose:

Project Backup

VGA:	PROJECT menu
VS:	SHIFT + PROJECT (F1)
VS-2480:	PROJECT

- Loading as the current project: →Project Load (p. 167), →Project Select (p. 169)
- Saving after each Operation: →Project Store (p. 169)
- Creating a new project: →Project New (p. 168)
- Saving name entries and commentaries: →Project Name (p. 168)
- Creating a backup copy on a CD-R: →Project Backup (p. 166)
- Recovering a project from a CD-R: →Project Recover (p. 169)
- Copying within one or to another →Partition: →Project Copy (p. 166)
- Deleting a project from the hard disk: →Project Erase (p. 167)
- Erasing overwritten recordings: →Project Optimize (p. 168)
- Protecting the project against recording and editing: →Project Protect (p. 169)
- Using a project in another VS model: →Project List (p. 167)
- Combining two projects to make one project: →Project Combine (p. 166)
- Copying one project with chosen tracks: →Project Split (p. 169)

Project Backup

All the →Project data is saved in a B. In this way, after a recover operation in the VS using →Project Recover, the initial state including all audio traces and settings can be re-established. In addition to the →CD Backup, projects in the VS-2480 can be saved to an SCSI hard disc and in the model VS-2480DVD additionally to a DVD (→DVD Backup).

Project Combine

Tracks of any project can be integrated in the current project. The operation "Project Combine" positions all tracks from the source project at the end of the present project (Destination) and subsequently deletes the entire source project from the hard disk. Through this the current project increases in size.

Application:
- Mixing numerous songs in one project with the same settings.
- Creating a CD of songs, which have been recorded in different projects.

Hint:
The project to be combined must be present in the current →Partition. If applicable this can be copied to the desired partition with →Project Copy.

SHIFT + Project → Page 2/4 → F Combine

The target project is always the current project. The source project is combined at the end of the target project. The VS defines a project's end with audio data and Markers. Only one source project can be attached in one procedure.

Fig.: Attaching a Project (Combine)

Audio data of active and passive →V-Tracks will be attached. By contrast, mixer settings, effects, locators, markers, scenes, automix data etc. can not be transferred to the current project.

Hint
- The attached project will be erased after the operation. If applicable, a copy should be made beforehand (→Project Copy).
- Chosen tracks can be better copied to a new project using →Project Split.

Project Copy

A physical 1:1 copy of the marked projects from the project list is generated by the operation P. The copy destination can be the current →Partition or any partition of the internal hard disk (VS-2480: also an external →SCSI hard disk). Creating backup copies and →Templates are applications for "Copy".

SHIFT + Project → Page 2/4 → Mark → F1 Copy

The target partition and the remaining memory storage space of this partition is shown in the destination field. To choose another partition, call up the page "Select Drive" via F4 SelDrv and then select the chosen partition from the list. YES chooses the partition to be the destination drive and returns back to the copy page. With F4 Copy, the operation is carried out, which can take a certain amount of time.

Tip
When hard disk problems ("Drive Busy") appear due to an overly intense →Fragmentation (among other things), copying the project to another →Partition often helps. The project can be saved using →CD Backup before further hard disk operations need to be carried out.

Project Erase

To delete →Projects, which are no longer needed, from the hard disk the function "Erase" is used. The memory space made available through this operation is then free to be used without any restrictions for other projects.

SHIFT + Project → Page 2/4 → Mark → F2 Erase

This operation is irreversible. An entire →Partition can be more simply erased using →ClrPrt.

Project Export

VS projects can be edited further not only in the VS models of the current series but also in the previous models. To do this, create a backup CD in the corresponding format. Be careful of the model dependent limitations regarding the number of tracks and the supported →Sample Rates and →Record Modes.

VGA:	Project → Export
VS:	SHIFT + Project → Page 3/4 → Mark → F4 Export

Project Import

Projects and songs can be loaded onto all* VS models via a CD-R. Keep the limitations in mind regarding tracks, effects, aux busses, the number of locators, the →Record Mode, →Dynamics etc. The procedure is practically identical to that of →CD-RW:

PROJEKT → "ATAPI" → LIST → MARK → PAGE → F3 Import → YES

Firstly convert VS-840 data to the VS-880 format.

Project List

The P. denotes the table of contents for the projects of one →Partition. The Protect Status (→Project Protect), →Record Mode, →Sample Rate, Name (→Name Entry), size, form and time of production can be read here. Generally to carry out an operation, a project must be appropriately marked (→Mark).

Project Load

In a hard disk recorder, audio data isn't loaded according to its mode of operation but rather, is read from the hard disk in realtime or written to the hard disk. The remaining time till the VS is ready for operation is determined by the respective settings data of a project, which are actually to be loaded from the working memory storage of the VS. In the project menu, the desired project in the active →Partition can be chosen from the →Project List (→Project Select) and can be applied after the security prompt (→Store Cur rent?) is confirmed.

VGA:	Project → Project List
VS:	SHIFT + PROJECT (F1)
VS-2480:	PROJECT

Loading a Project from the current Partition
The listed project names are only shown for the current partition. An inverted displayed row refers to the current project.

1. Clicking on the desired project positions the cursor (here in the form of a frame and arrow). This operation can also be carried out with the dial or the cursor button.

Project Name

2. Clicking or pressing F1 Select calls up a security prompt: "Select Project, Sure?", which is to be confirmed with YES. If "Select" isn't displayed for F1, press PAGE.
3. A security prompt referring to saving appears: "Store Current?" Generally the current project is saved with YES When changes carried out are *not* to be saved, choose NO.

The chosen project is applied after saving the current project and the respective →HOME page then appears.

Loading Projects from different Partitions
The VS hard disk is generally subdivided in →Partitions of 10 MBs. In total each partition can save 200 projects.

1. Call up the Project List (see above).
2. Move the cursor to the desired partition. These are labelled with "IDE:x".
3. Clicking or pressing F6 List calls up a security prompt:
"Change Current Drive to IDE... Sure ?", which is to be confirmed with YES. A further prompt referring to saving then appears: "Store Current?" (see above).

The project list with the selected project partition will be called up. From here the desired project will be loaded as per the description.

Project Name

Names and commentaries can be assigned when creating a →New Project and also later in the →Project List using "Name" (→Name Entry). The VS-2400CD shows the commentaries in the VGA monitor's →Screen Saver.

Project New

Generally for a new recording session, a separate →Project will be created. Its →Sample Rate and →Record Mode irreversibly establish its compatibility to other digital devices, the dynamic resolution, the number of recording and play back tracks and the maximum recording time.

Record Mode	VS-24xx to 48 KHz	VS-2000	VS-24xx > 48 KHz
M24	16 / 16	8 / 8	8 / 8
MTP	16 / 24	-	8 / 12
CDR	16 / 16	-	not suggested
M16	16 / 16	8/18	8 / 8
MT1	16 / 24	-	8 / 12
MT2	16 / 24	-	8 / 12
LIV	16 / 24	-	18 / 12
LV2	16 / 24	-	8 / 12

Recording /Playback Tracks at 44,1 KHz

The settings of the current project can be transferred with →Copy Mixer Parameter and →Copy Utility Parameter to the new project. It is recommended to give the project a meaningful name and an →Icon.

SHIFT + F1 (Proj) → F1 (New)

Tip:

- Similarly arranged titles can by all means be consecutively recorded in one single project. This simplifies not only the recording process because of corresponding settings, routings, monitor mixes, track names etc., but also the →Mix and the CD production.
- Self produced →Templates with prepared routings, track markings, mixer synchronization settings, midi settings, metronome settings etc. can be easily created with project COPY (→Project).

Project Optimize

P. erases all physical audio data, which still remains on the hard disk after multiple re-recordings of passages (→Take), within the current project and essentially could be reactivated using →Undo. This generally results in a considerable reduction of the project memory storage required on the hard disk. O. furthermore erases →Events for edit operations which have already been carried out. O. represents an irreversible function and is therefore to be carried out with due caution. Projects can be protected against operations that are accidentally carried out. To do this, the function →Project Protect found in the →Project List is carried out.

Tip
Before running a →CD Backup, Optimize should be carried out. This can often reduce the data to be saved.

Project Parameters
In the project page of the →UTILTY Menu in the VS-24xx, the following project oriented parameters are set:
- Digital Master/Slave clock:
  →Master Clock (p. 111)
- Outputting digital signals with reduced word length: →Dither (p. 58)
- Activating SCMS:
  →Digital Copy Protect (p. 55)
- Choosing a digital input:
  →Digital In Select (p. 55)
- Synchronizations Offset for Slave:
  →Display Offset Time (p. 57)
- Absolute/Relative time for the Sync-Slave:
  →Time Display Format (p. 225)
- Freezing a peak level for the display:
  →Peak Hold Sw (p. 145)

In the VS-2000, these Parameters are to be found in the menus →Digital I/O Parameter and →Display Parameter.

Project Protect
Using P., a project can be protected against accidental overwriting, editing, undoing and deletion. It still lets Locators/Markers be generated and scene as well as tempo map operations etc. be carried out. Saving these changes however, is not possible. To unlock a protected project, the procedure is repeated. Graphically, the protect status is symbolized in the playlist and the project list (→Project) with a lock icon.

VGA: Project → Protect
VS: SHIFT + Project (F1) → Page 1/4 → Protect

Only the currently opened project can be locked/unlocked. P. changes the lock status. This is marked in the following way:

Lock: Protect   off → on
Unlock: Protect   on → off

Project Recover
→CD-RW (p. 41)

Project Select
By clicking or pushing F1 Select in the →Project List, the chosen project will be loaded after confirmation of certain security prompts (→Project Load).

Project Split
The project operation "Split" is similar to the conventional copying of a project (→Project Copy). Audio Tracks, Mixer settings, Markers, Locators, Scenes, Automix Data etc. are also copied here. The difference between them is that here, the tracks to be transferred to the new project can be freely chosen. In this way, a customized copy of a project can be created. E.g creating a copy with all the settings but only the drum track.

VGA: Project → Split
VS: SHIFT + Project → Page 32/4 →
     Mark → F3 Split

A table of all V tracks of the current project appears. Played tracks are marked with a black box and can be selected with a mouse click or by positioning and pressing F4 Mark. After a security prompt, the new project complete with the selected tracks is created.

Project Store
Saving recorded data is one of the most important operations carried out by a digital worksta-

Project Top / End

tion. It should be done in regular intervals after recording and other important operations. In the VS, this procedure can be carried out either as a menu operation or directly on the user interface. Folders don't need to be chosen and paths don't need to be paid attention to. Instead, the project is overwritten at its original position in the project list with the current settings.

> SHIFT + ZERO
> (STORE)

Project Top / End

Using a button combination, the first or last recording positions can be directly moved to independently of when and in which track the first/last recording took place. For this the VS uses the active V tracks, which are presented in the form of bars in the →Playlist.

| First recording position: | SHIFT + FF |
| Last recording position: | SHIFT + REW |

Punch In/Out

P. is a special recording technique, where recording is only activated in the desired passage while it is being played back. After preparing the respective tracks (→Record Standby), start playback. Shortly before the desired passage, recording can be activated with REC and the corresponding part can be played again. After the passage, repeated pressing of REC swaps back to playback or STOP stops the recording.

Tip

- Punch In/Out can be carried out with a →Foot Switch or automatically (→A.PUNCH).
- Via a synchronized →MIDI Sequencer, this function also can be carried out.

Q

Q

The filter quality (Q) signifies the width of the effect range of a filter in the parametric →Equalizer. High values represent narrow frequency ranges. These narrow ranges mean that the neighboring frequency ranges both right and left of the center frequency are affected less.

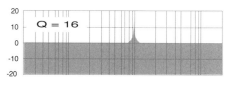

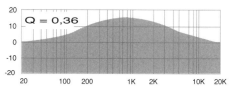

Fig.: Filter Quality Q

"Q" represents a non dimensional unit, which is measured at 3 dB both right and left below the level of the middle frequency. With a value of 'Q=1,4', approximately one octave above and below the set frequency will be affected (e.g. with 1 KHz from approx. 500 Hz to approx. 2 kHz). More octaves are affected when the quality is lower (a broad frequency band). Extremely high quality values mean that almost only the set frequency will be controlled (e.g. one tone). For a more musical characterization of the frequency bandwidth, the parameter →Bandwidth is used. (This is not to be confused with Q!). High Values here refer to a broad frequency band. The following table should make this clear:

Q	0,4	0,67	1,41	2,87	4,8	7,21	14,4	16
BW	3	2	1	0,5	0,3	0,2	0,1	0,09

Quantization

Q. is the bit rate (word length) used in the →A/D Conversion. It also definitively specifies the →Dynamic of a digital audio signal. The conventional 16-Bit (CD standard) is being replaced more and more with a bit rate of 24-Bits. In addition to the word length, the →Sample Rate is significantly responsible for the digital audio quality.

Quantize

Within the operation →Phrase Copy when the function "Quantize" is activated, the VS automatically corrects the TO time points in a quarter pattern. The condition for this is multiple copies with a "Copy Time > 2", whereby the rhythmical structure stays the same particularly for drum loops or in connection with a synchronized midi sequencer. Quantizing the TO points will also be correctly carried out when the tempo is changed (→Tempo Map).

Tip

For the Sound design, the effect →Lo-Fi Processor (p. 106) allows to decrease the Bit rate to achieve a Lo-fi Sound.

Quick Routing

A quickly routing of recording tracks to input channels can be carried out directly with the Status/Select button on the VS. This method is especially suited to the VS-2000 because of its permanent fader allocation. The starting point here is always from the track that is to be recorded. By pressing and holding one of the STATUS buttons, the "Quick Routing" menu is opened*. The virtual cable connection is routed with SELECT (or INPUT in the VS-2000) to an input channel.

* The response time for calling up the menu is specified by the parameter →Switching Time.

Details referring to quick routings can be found under "→Quick Routing Via Status/Select Buttons (p. 75)" in →EZ Routing (p. 74).

R

R-4 (Pro)

The R-4 made by Roland's subsidiary company Edirol, is a portable 4 track recorder, which provides an internal 40 GB hard disk. Through its optional battery operation, four integrated →Microphone Preamps with analog limiters, and switchable phantom power, it is particularly suitable for recording on location with 1 (Reporting) to 4 simultaneous track recordings (→Surround).

Fig.: Field Recorder R-4

The →Sample Rates 44,1; 48 and 96 KHz and the →Record Modes 16 and 24 bit in wave and broadcast wave format are available for projects. The recorded audio data can be graphically edited in the internal display. Additionally, the audio data can be exported to a computer via USB or compact flash cards. The R-4 is fitted with a LANC connection. This means that it can therefore be externally controlled directly by and perfectly in frame sync with a digital camcorder. The VS models need an additional synchronizer →SI-80SP for this. The "R-4 Pro" can be synchronized via →SMPTE Two microphones and both the loudspeakers are integrated economically in the housing of the device. For further editing, R-4 tracks can be used in the VS as follows:

R-Bus

USB:
Transfer tracks via USB from the R-4 to a computer. Burn the audio data in wave format onto a CD-R. In the VS, this CD can be used via →Wave Import.

S/PDIF (Pro version: AES/EBU):
→Recording of digital Signals (p. 181)

R-Bus
This digital →Interface developed by Roland makes transferring digital audio data from one R-Bus device to another possible. Through this, eight audio signals (max. 24-bit/96KHz) can be simultaneously sent and received. Midi control data and the →Clock are additionally transferred. The following devices provide an R-Bus connection:

- VS-2480 (2x), VS-2400, VSR-880
- →AE-7000 (AES/EBU-Interface)
- →ADA-7000 (Analog-Interface)
- →DIF/AT, →DIF-AT24 (Adat-Interface)
- →RPC-1 (PCI-Card)
- →V-Fire (Firewire-Interface)
- VM-7000, VM-3100Pro (Digital mixer)
- XV-5080 (Sound-Module)

The analog interface ADA-7000 can be used effectively particularly for the VS-2400. This exclusively enables the simultaneous recording of 16 microphone or line signals (→Balanced) possible. Eight additional analog outputs for transmitting bus and individual signals are available.

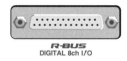

Fig.: R-Bus Socket

In the following example, the use of an analog effect device via the →ADA-7000 is described. This also exemplarily describes the connection of inputs and outputs.

External Effect Processor via ADA-7000
The external processor (→Send/Return Effect, e.g. Reverb processor) should be fed with a mono signal via →AUX Bus 8. The VS' default routings should be maintained for this example:

1. Connections and Powering Up

1.1. R-Bus
When turned off, connect the VS and the ADA-7000 using an R-Bus cable. For the ADA to be recognized in the VS, the following Powering Up sequence is to be adhered to:

ADA ➔ VS

1.2. ADA-7000 - External Effect Device
ADA-Output 8 ➔ Mono Input Effect
Effect Outputs ➔ ADA Input 7/8

2. Configuring the ADA-7000
The VS with its internal master clock will be used as the →Clock-Master although principally the ADA can also be used for synchronization and time measurement. Controlling the level of the input signal is directly carried out with the Sens knobs of the ADA because of the level meter on the VS that is to be controlled.

Utility ➔ R-Bus Config Remote Control = On ➔ F1

Clock Source: R-Bus
Sampling Freq: like a Project (mostly 44,1K)
Input Sens 7/8: -10 dB (temporarily)
Front Panel Control: On

3. Checking via the signal tone generator
Using the →Generator/Oscillator the routing and the input and output levels can be controlled with →Level Meter.

Utility ➔ Gen/Osc

- With a level of 0 dB (1KHz) a test tone is outputted via the Generator's AUX8 Send knob into the Aux bus 8. Then, return back via HOME.

In the level meters AUX/DIR and OUTPUT (R-Bus), -12 dB will be respectively displayed for the corresponding bus.

- In the effect device, the test tone can be used to set the correct input level.
- Level meter INPUT of the VS displays the corresponding output level of the effect device.
- In the ADA, both signals are regulated with the knobs 7 and 8.
- With the VS' default routing, the external device's FX returns are fed into inputs 15/16 of the →Input Mixer. These can be →Linked.

4. Leveling Aux sends in Mixer Channels
Deactivate the Generator/Oscillator. In the respective mixer channels, the desired part of the signal that is to be sent to the external effect device is leveled via aux send 8 according to musical and aesthetical guidelines.

5. Variations
For reasons of clarity, the default routings are used in the described example. For sending signals to the ADA however, any busses or also →Direct Path can be used for one's own applications. (→Output Assign).

External Effect Device as an Insert
To use an Insert, principally the same connections and routings are established. To conserve a flexible Aux Bus, a Direct Path is used at least to send the individual signal. Deactivate the Mix switch in the corresponding channel.

Two VS-Recorders
The following example describes overwriting audio data and the synchronization of two VS Recorders (not in the VS-2000). Connections with an R-Bus cable are to be undertaken while the machines are switched off.

1. Audio Transfer
1.1. Transmitter (VS1)
The VS, which transmits the audio signals, provides the digital clock via it's internal →Master Clock. The desired tracks are routed in the page Output Assign of →EZ Routing either with →Track Direct Out, →AUX Bus or →Direct Path to any R-Bus outputs.

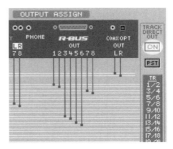

Fig.: Direct Outputting of Tracks via R-Bus

The diagram above shows an example of routing in the mode "Track Direct Out". The routing is achieved as follows:

Track	1/2	11/12	13/14	3/4
R-Bus	1/2	3/4	5/6	7/8

1.2. Receiver (VS2)
In the receiver VS, the Master Clock parameter needs to be set to "R-Bus". Signals, which are fed into the R-Bus input, can be routed in the EZ Routing's →Patchbay (→Recording of digital Signals (p. 181)). In the default settings, the R-Bus inputs are patched to the mixer channels 9-16. Any routings can also be carried out here.In this example, the track signals 1/2 of the sending VS1 are transferred to inputs 9/10 of the receiving VS2.

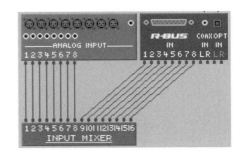

Fig.: Patchbay in the VS-2400

Track VS1	1/2	11/12	13/14	3/4
R-Bus	1/2	3/4	5/6	7/8
Input VS2	9/10	11/12	13/14	15/16

R-Bus Patching

2. Synchronization
If synchronization of both VSs is necessary, the following settings are to be carried out (→Synchronization):

Sync-Parameter	Master	Slave
Sync Mode	INT	EXT
R-Bus Sync Gen	MTC	-
Frame Rate	same as Slave	Same as Master
Ext Sync Source	-	R-Bus

The Midi settings used for transferring the transport commands can be taken from the following table:

Midi-Parameter	Master	Slave
SysEx. Tx Sw	On	-
SysEx. Rx Sw	-	On
MMC Mode	Master	Slave
MMC Source	-	R-Bus

In addition to the transport functions, the REC button will also be externally controlled using MMC.

R-Bus Patching
There are interesting possibilities for unusual routings in the VS-2480 because of its two separate R-Bus connectors. These are made possible by connecting both sockets with an R-Bus cable (!). The R-Bus 2 outputs are now fed into the Input Mixer's R-Bus 1.

Application
- Creating real →Subgroups
- Filtering Aux Sends with channel EQ for Live Monitors.

R-DAC
This method of coding for digital audio signals developed by Roland makes largely loss less saving possible in contrast to the data reducing methods like ARTAC, MP3 etc. In comparison to recordings in the 16 bit linear format, 50% of the necessary hard disk memory can be saved here, which means, there is 50% more →Recording Time possible. Using the R-DAC method, the audio data to be saved is compressed on the hard disk and then respectively decompressed for playback. In the models VS-24xx, M24, M16 and CDR represent non compressed →Record Modes. In the VS-2000 a data compression is entirely dispensed with.

Rate
Within →Effect editing the parameter "Rate" determines the speed or frequency of a modulation while "Depth" is responsible for the depth or intensity.

Ratio
In the →Dynamics the compression/expression ratio of input to output signals can be specified using R. This takes effect if the →Threshold level is reached or fallen below. With a value of 1:1 no level change will take place (linear characteristic line). This corresponds to a deactivation of the level amp. A ratio of 1:2 produces a halving of the volume, 1: ∞ a limiting/Gate function.

RCA
The →Unbalanced connection →Cinch is also called RCA.

Realtime, Automix
Within →Automix both dynamically flowing movements in realtime and the static snapshots can be used.

Record Bus
The transfer paths used to transmit the recording signals to the recording tracks are called Record Busses (→Fig.: Signal Flow when Recording (p. 177)). They are not visible to the user and appear solely in the form of the maximum tracks to be recorded (→Record Mode). Nonetheless R.s can be routed together independently of the →Track Mixer to the →Monitor Bus and correspondingly edited there.

Record Monitor

EZ-Routing → Output Assign:
Route MON to REC

In this case, the listening volume of all R.s is leveled with the REC BUS knob (see figure above).

Application

- Listening to the recording signals without further settings.
- →Master Insert with External Effect

Record Mode

When creating a new project, the R. irreversibly configures the word length (→Bit Rate) for saving the digital audio signals and therewith also the →Dynamics and the degree of compression (→R-DAC). The setting →MTP is recommended for the models VS-24xx, which represents a compressed 24 bit format (→16 bit / 24 bit). Dubbing tracks in order to create a CD from a CD, LP, MiniDisk etc. should be created in →CDR format to save time. The remaining Record Modes provide compatibility to the previous VS models (→Project Export) and help attain extremely long recording times. The choice of the Record Mode determines the number of available recording and playback tracks and the respective →Recording Time:

Record Mode	VS-24xx 44,1 KHz	VS-2000	VS-24xx > 48 KHz
M24	16 / 16	8 / 12	8 / 8
MTP	16 / 24	-	8 / 12
CDR	16 / 16	-	illogical
M16	16 / 16	8/18	8 / 8
MT1	16 / 24	-	8 / 12
MT2	16 / 24	-	8 / 12
LIV	16 / 24	-	18 / 12
LV2	16 / 24	-	8 / 12

Tab: Recording / Playback Tracks

Hint
The record mode (word length) is irreversibly set at the creation of a new project. It cannot be subsequently changed.

VS-2000
Here only the 16-bit and the 24-bit linear recording modes are provided. For recordings with numerous instruments and a normal dynamic, the 16-Bit is the format to choose. In contrast, for extremely dynamic solo recordings (Concert guitar, vocal solos and also piano trios), using the qualitative high value 24 bit is advised. This accompanies an increase in the necessary hard disk storage space.

Record Monitor

When playing back armed tracks, the R. parameter determines which signals are heard: Either the routed input signal or the already recorded track signal.

Utility → PlayRec1 → Record Monitor

1. Source:
In play and stop mode, the respective input signal is always heard which lets the recording be simulated (→Rehearsal).

2. Auto:
When the project is stopped, the input signal to be recorded can be heard via the Track Mixer. By contrast, when PLAYing, monitoring changes automatically to the recorded material (this describes normal playback).

Manually Changing:
Within the operation mode "Auto", changing the signals to be heard can also be manually carried out: This applies to the Input signal and already recorded signals. To do this merely push the blinking, red status button again until it changes to orange/red and through this outputs the input signal. This input status then stays the same till the project is stopped. Pushing the button again toggles back to the recorded signal (same as for playback). By doing this, not only level settings for →Punch In/Out, but

Record Standby

also play modes and articulation of already recorded passages can be compared to passages that are to be replaced. In this way for new recordings during playback the corresponding input channels can be easily listened to.

Record Standby   ▮▮▮
R. describes modes of operation whereby a track is already armed but recording doesn't take place. The corresponding STATUS button on the VS blinks red, in the LCD the respective track numbers are inverted and in the VGA display, play is marked red. R. can be used just as well in stop and play modes. This enables hearing the input and an already recorded signal to hear if any changes should be made.

Stop:
The routed input signal is played

Play:
According to the setting of the parameter →Record Monitor, either the already recorded signal or that routed input signal will be heard.

After the desired tracks have been armed, press the REC button when the project is stopped. The standby status is then shown by the REC button blinking. The recording of Audio tracks (→Recording) or in →Automix the recording of automix data only begins after subsequently pressing PLAY. The VS STATUS button then blinks red. In the VGA the indicator for the corresponding track changes from PLY (yellow) to REC (red).

Recorder
Recording device for analog or digital audio data.

Recording   ▮▮▮
Within music production, R. represents an essential part of the process. Whilst →Editing, the →Mix and →Mastering generally aren't carried out under any time constraints and operations can be repeated any number of times, there are boundaries when recording. In addition to the technical settings, this concerns in particular subjective factors such as impatience, being unsure and stress, felt not only by the musicians who are playing but also but the sound technicians themselves. Competent mastery of the VS's recording process and thorough preparation is therefore recommended.

1. New Project
Generally a new project (→Project New) will be applied to the recording. Attention should be paid particularly to the →Sample Rate and the →Record Mode although the default parameters set represent sensible settings:

SHIFT + F1 (PROJ) → F1 (NEW)

Practical settings for «New Project»	
VS-24xx:	44,1 KHz / MTP
VS-2000:	16 Bit

Settings of the current project can be transferred to the new project using →Copy Mixer Parameter and →Utility → System Parameter. It is recommended to give the projects meaningful names and as the case may be →Icons.

Tip:

- Similarly arranged tracks can be recorded consecutively in a single project. This simplifies the recording process because of corresponding →Dynamics and →Equalizer Settings, Routings, Monitoring, Track names etc. and also the →Mix and creation of CDs.

- Personally produced →Templates with prepared Routings, Track identifications, mixer synchronizations, Midi and metronome settings etc. can be created easily using →Project Copy.

2. Connections
The desired instruments or microphones are connected using appropriate connection cables with the corresponding input jacks. In the beginning it is recommended to use the input

Recording

jacks, which correspond to the desired recording tracks: Input 1 'Track 1', Input 2 'Track 2' etc. This is always the case after creating a new project or after the →Parameter Initialize and saves →Routing for the time being. Recording digital input signals requires special presetting and routing (→Recording of digital Signals).

Phantom Power:
Only →Condenser Microphones and special →D.I. Boxes require →Phantom Power of 48 V. for operation. Whilst the VS-2000 provides hardware switches on the back panel (1/2, 3/4, 5/6 and 7/8), the VS-24xx carries out the settings for each input jack via Menu:

SHIFT + EZ ROUTING

Phantom Power can also be connected to the routed input in the according input channel of the VGA.

3. Signal Flow when Recording
Input signals firstly pass through the →Input Mixer before the hard disk recording. Signal editing carried out here (Volume, EQ, Insert Effects etc.) directly and irreversibly influences the waveform to be recorded. As can be inferred from the block diagram, listening to the input signal during recording takes place only in the →Track Mixer. The analog input signal (microphone) is firstly boosted in an analog pre amp to the necessary level while the analog/digital conversion occurs in the dedicated →A/D Converter.

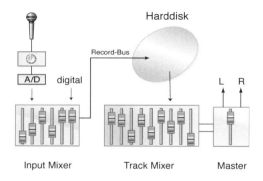

Fig.: Signal Flow when Recording

4. Level Setting of analog Input Signals
To achieve a correct level, the input signal is to undergo an inspection in the editing stage. The following diagram shows the measuring points available in an input channel.

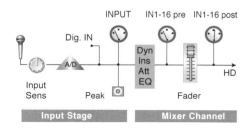

Fig.: Level Measuring Point in the Input Mixer

Normally, unedited signals are recorded. This limits leveling to the analog pre amps. Although seldom the case, if signal editing is carried out during recording, further level checking is recommended (→5. Recording with Effects). Parameter settings that have been undertaken in the Input Mixer go into the recording tracks as data and can't normally be undone again. The following is a short guide for controlling analog input signals without editing (linear). A detailed explanation for the steps to be carried out are described in the following sections.

Level Setting in the Input Mixer

1. Set the corresponding Fader in the Input Mixer to 0 dB.
2. Call up the Level Meter IN1-xx.
3. Control peaks to a max. of -4 dB with the Input Sens knob.

4.1 Leveling the Analog Input Stage
Using the →Sensitivity Knobs (Gain) the analog input signal will be adjusted to match the necessary level of the pre amp.

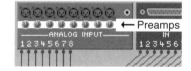

Fig.: Preamps in the EZ Routing Disp

Recording

This is monitored here by →Level Meter INPUT, which allows the analog section before the mixer channel to be watched. This level meter is able to show both analog and digital input signal levels and is called up in the following way:

VS-Display	VGA
F1 → INPUT Press Page, if necessary.	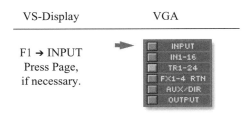

Fig.: Calling up the Input Section's Level Meters

To achieve a large →Dynamic and an optimal →Resolution, high recording levels should be strived for. →Peaks should seldom be allowed to cross over the preset →Input Peak Level of -6 dB and never over the full level control of 0 dB.

INPUT PEAK:
INPUT →Peak Indicators are to be found at the same measuring point as the Input Level Meters. These show peaks in the input stage *before* the Input Mixer and solely indicate when the warning level is exceeded. With the default -6 dB enough →Headroom is available so that if the INPUT PEAKs occasionally blink it can be assumed that the level has been correctly set.

4.2 Leveling the Input Mixers
The input signal always flows through the corresponding channel of the Input Mixer. (see "Fig.: Signal Flow when Recording" p. 177.).

VS-Display	VGA
F2 → IN MIX Press Page, if necessary.	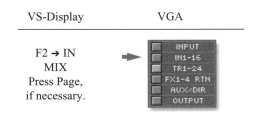

Fig.: Calling up the Input Channel's Level Meters

Generally only the channel volume is to be set with the Fader to 0 dB. Only if necessary can sound changes already be carried out for the recording (→5. Recording with Effects (p. 179)). The Level Meters "IN1-xx" are available to control the levels in the input mixer channels (see "Fig.: Level Measuring Point in the Input Mixer" p. 177.). This can be take place either before or after the Fader (→Pre/Post). The Level Meters IN1-xx can be neglected in the case of linear recording without the use of EQ, Dynamic and Effect Insert and with a →Level of 0 dB in the Input Channel. This results in:

1. No level boosting or attenuating
2. The level values of the Level Meter "INPUT" and "IN1-xx" are identical.
3. The level indicators for →Pre/Post in the Level Meters IN1-xx are not different.

In general, the following then applies:

Recording settings in the Input Mixer
Set the Channel Fader in the Input Mixer to **0 dB**

The 0 dB setting of the Mixer Channels for the Input Mixer is achieved as follows:

VS-24xx:
In the VS' «FADER» section, firstly press the →Fader Buttons IN 1-xx. The Faders are then assigned to the first section of the Input Mixer. The FADER section is found above the red Master Fader. The Fader buttons assign the "Universal Fader Unit" to the respective mixer section.

Tip:
The 0 dB position of a Fader can be automatically called up using CLEAR + CH EDIT in the corresponding Mixer Channel.

VS-2000:
The eight analog and the digital input sockets are permanently connected to the Input Mixer channels - The EZ Routing's Patchbay only provides clarity. No physical Faders are avail-

Recording

able in the Input Mixer. Using one of the round buttons in the INPUT section calls up the desired →CH EDIT page. After positioning the cursor over the Fader field (... dB), settings can be carried out with the Time/Value dial.

Control with the mouse and menu
The desired Input Mixer Channel (IN xx) can be called up using Mouse and Menu Control in the VS Display and also in the VGA as follows:

    Mixer → Ch View → Input Mixer → IN xx

Deviation from 0 dB Settings:
For certain applications, it is necessary to deviate from the rule, "*Set all Faders to 0 dB in the Input Mixer when recording.*" This concerns previously established fades and level nuances and the balancing of EQ, Dynamics and Insert settings in the Input Mixer. In the same way and for musical reasons, premixing numerous input channels to *one* Track (e.g. recording multiple choir voices to one Mono/Stereo track) requires changing the volume with Fader in the Input Mixer (see Diagram).

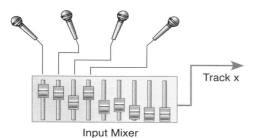

Fig.: Pre Mix in the Input Mixer to record numerous signals to one Track

5. Recording with Effects

In some applications linear input signals won't be recorded. Instead, previously edited signals and →Effects will be recorded. This particularly concerns style definitive Insert effects such as Guitar amp Modelling, Vocoder, specially timed Delays etc. but also previously established filters, dynamic settings and volume fading/gradation. In order to be able to record audio data changed with →Insert, editing the signal takes place in the Input Mixer located before the hard disk (see "Fig.: Signal Flow when Recording" p. 177.). It is not possible to restore the original signal after recording. Due to →Effect Economy, →Send/Return Effects such as Hall, Delay, Chorus etc. can also be recorded together with the original signal or to separate tracks. In order to do this the relative →Effect Return signal is to be routed to the desired recording track in →EZ Routing.

Accomplishment:
The settings are put into effect in the relative Input Channel which is connected to the corresponding input jack. The following →Routing represents the default setting:

Input 1 → Input Mixer Channel 1
Input 2 → Input Mixer Channel 2 etc.
Diverse methods are available to call up an Input Mixer Channel:

Controlling with the Mouse and Menu
An Input Mixer Channel (IN xx) can be called up similarly via the Mixer Menu:

    Mixer → Ch View → Input Mixer → IN xx

Button on the VS-24xx:

- Press IN1-xx (→Fader Buttons*).
  This assigns the CH EDIT buttons to the first section of the Input Mixer.

- After pressing the corresponding CH EDIT buttons, the respective →Channel Strip is called up.

  * *The FADER field is found next to the red Master Fader.*

Tip:
Confusion with the Track Channels or the second section of the Input Mixer arises often. This is mostly the result of wrongly selecting the FADER buttons. To make sure that the correct Track Channel or section is being used, pay attention to the Name field on the left at the top of the current Mixer Channel (e.g "IN 8").

Recording

Buttons on the VS-2000:
The input sockets here are permanently assigned to the respective Mixer Channels. The →Patchbay serves merely for visual monitoring. To call up the desired channel strip, press the corresponding button in the INPUT section. Confusion of the Track Channels shouldn't arise here due to the permanent assignment of the buttons.

Controlling the Signal Level:
When Input Mixer Channel's Parameters have been changed, the signal level in the Input Mixer needs to be checked. For this the level meter "IN1-xx" is used. As can be seen in the following diagram, volume changes can appear through using →Dynamics and insert effects although the input signal at the Level Meter INPUT shows a correct level. The necessary level corrections can be accomplished using the Faders and the →Attenuator, which allows an additional volume change.

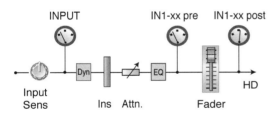

Fig.: Level control during signal editing.

6. Routing from Inputs to Tracks

Establishing a signal connection from a source to a desired target using patch cables is called →Routing. The virtual "Cables" used in the VS serve to visualize routings in addition to the incorporation of the connection. When recording, the signal path flows from the input jacks over the corresponding Channels of the Input Mixer to the Recording Tracks (see "Fig.: Signal Flow when Recording" p. 177.). In the following diagram the analog input jack 1 is routed to the channel's 1 input of the Input Mixer. The signal path then flows from the output of channel 1 of the Input Mixer to Recording Track 1.

Limitations:
- Routing a channel to itself is not possible.
- Routings can only be carried out when the recorder is stopped*[1].

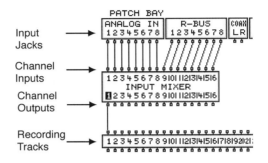

Fig.: Routing with a virtual cable (VS-2400)

Tip
The default Settings can be reverted to for recording the first eight tracks*[2]. Personally configured routing is not necessarily essential. The input jacks 1 - 8 are already assigned to the Recording Tracks 1 - 8 so that the microphone for the recording of a desired track merely has to be connected to the jack with the same number. For example; Input jack 8 → Input channel 8 → Track 8 etc.

[1] Routings via →Direct Path allow changing the recording source during recording.

[2] VS-2480: sixteen Tracks

6.1 Patchbay → Input Mixer

The patchbay (not in the VS-2000) is located before the Input Mixer and establishes the connection of the Input jacks to the internal inputs and Input Mixer Channels. It is essential because there are fewer input jacks than Mixer Channels in the VS and the choice of input takes place in this way. The VS Patchbay already contains the →Microphone Preamp and the →A/D Converter. This is why only digital audio signals are routed from here. The connection occurs in pairs.

6.2 Input Mixer → Recording Track

To simply record the first eight channels (16 in the VS-2480), routing is not necessarily essential. Simply re-plugging the cables into the Input jacks of the corresponding Tracks also leads to the desired result. Routing is however necessary for the following applications:

- Recording to Recording Tracks 9-xx (VS-2480:17 - 24).
- Recording via the Hi-Z guitar input to desired tracks.
- Recording with identical settings of an input channel to further tracks.

The further procedures are available:

Quick Routing via buttons.
1. Hold the STATUS button for the Target Track
2. Change to the Input Mixer and press the desired CH EDIT button for the Input channel (→Quick Routing (p. 171)).

"Track Channel Edit" Page
1. In the Track Mixer, open the CH EDIT page of the target track.
2. Press F2 (Assign)
3. Position the Cursor at the desired input field and mark it with the dial.

EZ-Routing – Virtual Patch Cable:
1. In Input Mixer in the bottom row "Track Assign" select the desired input channel.
2. Route using the dial to a Recording Track (Alternatively: drag with the mouse)

6.3 Bouncing
Internally recording tracks to further tracks is called →Bouncing. Furthermore the following are used as the source: →FX Return Channel, →AUX Bus/→Direct Path Master and →Generator/Oscillator.

7. Arming the Recording Tracks
Tracks which are to be recorded are armed with their STATUS button ("Record Standby", →Track Status). Input channels routed to armed Recording Tracks can now only be listened via the Track Mixer when the project is stopped. (The respective Input Channels are automatically disconnected from the Mix Bus in the Input Mixer). For a →Rehearsal run in order to be able to play alone to recorded tracks, the input signals must also be audible in the playback in the Track Mixer. This is made possible with the parameter →Record Monitor.

8. Monitoring
Because oftentimes a separate recording room is not available, for acoustic recordings listening occurs with →PHONES and not over the →Studio Monitors. A special →Headphone Mix for the musicians, independent of the normal Stereo Mix can be created via →AUX Bus. For projects with a defined tempo, the integrated →Metronome or the sounds of a synchronized sequencer system can be used.

Talkback:
For communication between the Sound engineer and the musicians, an additional →Talkback microphone needs to be provided, which is exclusively fed into the →AUX Bus.

9. Recording
Through activating REC, record is armed (→Record Standby). Subsequent PLAY starts the recording of the armed tracks. Pressing REC again deactivates the recording process and changes back to Play Mode. This can overwrite or completely erase the target passages without interrupting playback. Activating/Deactivating Record via the STATUS button while recording is not possible. Writing over the entire recording or also chosen passages generates respectively a new →Take and leaves the Original recording untouched. →Undo enables returning to any editing step or recording.

Recording of digital Signals

The multitude of devices with digital connections existing on the market suggest using the digital inputs on the VS. The resulting advantages of this are represented in an unchanged signal quality and also the economical use of the VS's analog inputs.

Recording of digital Signals

Application:
- Dubbing of CDs, Dat, DAW etc.
- Using digital Effect devices
- Economical use of the VS inputs (→Talkback via an additional converter etc.)

To record digital audio signals the following procedures are to be carried out:

1. Connect the digital output of the external devices with the corresponding digital input on the VS.
2. Set the Master Clock
3. Route the digital input being used to the desired Mixer Channel (see "6. Routing from Inputs to Tracks").

Sample Rate and Record Mode
The →Sample Rate of the digital signals that are to be recorded must correspond to those of the current VS project. Otherwise the VS synchronizes to the external Clock and then changes the pitch automatically. Except for the VS-2000 which records audio material only with 44,1 KHz, the Sample Rates of 96, 64, 48, 44,1 and 32 KHz are available. Sample Rates deviating from these values can be produced using →Vary Pitch (not on the VS-2000) in a limited range. Digitally dubbing →Audio CDs requires a project Sample Rate of 44,1 KHz. A higher word length (→Record Mode) for the project than that of the input signal to be recorded is still to be strived for. Although the original signal's bit rate is maintained, the signal quality takes advantage of the higher precision whilst editing via →Equalizer and →Dynamics. For intended editing operations it's therefore advised to choose the 24-bit mode (→MTP). When creating a →New Project, the parameters "Sample Rate" and "Record Mode" are irreversibly configured.

Tip:
Although preference is to be given to digital recordings, due to the high resolution of the 24 bit →A/D Converter, the signal quality of analog recordings will not be audibly any worse.

Switching to analog recording when the Sample Rates don't correspond can be carried out without hesitation.

Digital Inputs
Digital audio signals are fed into the special →S/PDIF and →R-Bus connections on the VS. Using a Digital Interface (→AE-7000), additional →AES/EBUE inputs and outputs are available (not in the VS-2000). The particular VS models are equipped with digital connectors:

	VS-2480	VS-2400	VS-2000
S/PDIF coaxial	1	1*	1
S/PDIF optical	1	1*	-
R-Bus connector**	2	1	-

* only able to be used alternatively
** 8 IN//OUT per connector

Master Clock
To avoid clicks and artefacts when digitally transferring audio signals, the Sample Rates of both devices must be synchronized (→Clock). Using the simple connection, the second device (Slave) then adopts the Sample Rate of the first digital device (Master).

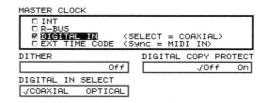

Fig.: Source for the Master Clock (VS-2400)

In the VS, changing from the internal to the external clock is carried out by assigning the respective digital inputs in the menu →Master Clock:

Utility → Project → Master Clock

Here the appointed Input used for incorporating the digital connection is chosen: Either →Coaxial or →Optical (In the VS-2400 with an additional SELECT field), →R-Bus or →Ext

Timecode (Timecode synchronization as Slave). When the clock is correctly in sync, the VS displays the message, *"Digital in Lock!"*. Wrong assignments of the Master clock and seldom defective cables (→Digital Cable) are mostly the cause of Clock problems and are shown with:

"Digital In Unlock, Change to Internal Clock?"

Hint:
Sample Rates which don't correspond are only tolerated by the VS-24xx models because of their implemented →Vary Pitch function. Always pay attention to the accompanying pitch changes of recorded tracks and also signals that are to be recorded.

VS-2000:
Access takes place here via the "DIGITAL" Utility menu. Due to the lack of optical and R-Bus inputs, solely DIGITAL IN can be selected.

Recording from a CD Player:
To digitally record from a CD player, the following settings are necessary in the VS:

1. A →Sample Rate of 44,1 KHz for the project (see above).
2. Deactivate the copy protection (SCMS)

For unlimited use of the VS, the copy protect flag (→SCMS) contained in the data stream can be deactivated using the parameter "CD Digital Record". This displays the conditions for digital recordings from a CD Player:

Utility → Global → CD Digital Record = ON

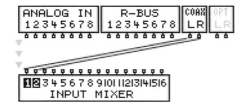

Fig.: Routing from COAX IN to Input Channels 1/2

R-Bus:
Up to eight digital audio signals can be transferred bi-directional using the Roland's own digital format →R-Bus. Devices which are already equipped with an integrated R-Bus connection don't need any further interfaces for transfers. To use →AES/EBU connections, an optional →AE-7000 is to be used. The necessary routing is to be carried out as written above.

Recording Time

The maximum recording time is determined by the number of Tracks, the size of the →Partition. the →Sample Rate and the →Record Mode. The maximum recording times of the respective VS models with a Sample Rate of 44,1 KHz and a Partition size of 10 GB can be taken from this table. When put into practice, they are underlying, sensibly applied Record Modes. The Values have been rounded.

Tracks	MTP	Liv2	16-bit	24-bit
1	72	143	36	23
2	36	72	9	12
8	9	18	$4\,^1/_2$	3
16	$4\,^1/_2$	9	$2\,^1/_4$	$1\,^1/_2$

Tab.: Recording time [h]
Darkened area: VS-2000

In contrast to tape oriented systems with a recording time dependent on the tape length, the utilization of memory for each Track on HD recorders is 'pool' oriented. A Recording here merely uses its necessary capacity, which is taken from the entire 'pool'. In doing this, it is irrelevant which →V-Tracks are being used. The available total time is the result of adding all the individual recording times together. Because each new recording of a passage generates a new →Take, this is also relevant to memory space. Using →Optimize, re-recorded passages can be erased from the hard disk and the relative memory space then becomes free

Recover

again. V-Tracks no longer needed must, however, be erased manually using →Erase and subsequently Optimized. Recordings can only be carried out in the current Partition. The available memory space can be read in the Remain Field, which can be found on the top right in the LCD and under the playlist in the VGA.

Utility → Global2* → Date/Remain Sw
→ Remain Display Type

Optionally, the remaining time, the capacity in MB or %, the remaining →Events or also the Date can be displayed (VS-2000: System).

Recover
→CD Recover (p. 40)

Red Book Standard
This standard, established by the firms Sony and Phillips, describes the configuration of a →CD Audio: The location and the mode of the CD tables of contents (TOC), the data format of the audio data, the →Finalize, the Timecode format, the max. number of 99 Tracks with a total play length of approx. 80 min. etc. When in →CD Burn using the mode →Disc at Once, the VS is able to create corresponding Red Book compatible CDs.

Tip:
Opposing Information saying otherwise, Duplication Companies also accept CD-R as a Premaster, which is created in the more comfortable →Track at Once procedure and which *doesn't* accord to the Red Book Standard.

Redo
REDO cancels the last →Undo operation carried out, if the project hasn't been saved yet. This returns the project to its status previous to the last Undo Operation and accesses all Undo Levels in the list:

VGA:	Click 🔁 (Redo)
Mouse:	Right click in the playlist
VS:	SHIFT + UNDO

Refresh Rate
This parameter determines how often the →VGA Monitor is to be refreshed:

Utility → System2 → Refresh Rate

R. determines the number of pictures to be configured each second. For a TFT Display owing to system dependent idleness, rates of 66 - 75 Hz are able to be chosen. This can differ from case to case according to picture quality. Generally these settings should be carried out particularly for tube monitors (CRT) using the Time/Value dial in order to easily be able to change back to the original value in the case of picture malfunction.

Region Arrange
Song Regions such as verses, refrains, Bridges etc. can be easily rearranged using the special copy operation "Region Arrange". To do this the Regions are firstly defined using Markers and subsequently grouped in the arrangement table and edited. The last step is to copy the new arrangement to any time position.

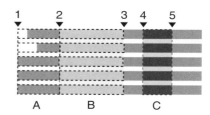

Fig.: Marker positioning for Region Arrange

Such Regions extend over all 24 active tracks in the playlist and their length is determined by two markers. In the diagram above, this concerns the Markers 1 and 2 for the Region A; 2 - 3 for Region B and 4 - 5 for Region C, respectively for all active V-Tracks.

Application:
For multiple rearrangements. In contrast to Region copy, Regions are maintained here at the Marker Positions.

Tip
In case multiple songs have been recorded in the project, their order can be easily changed using R.

1. Creating Markers for a Region
At the start and end of the relevant Region, →Markers are generated: →Positioning the project at the desired time (with the dial, transport buttons etc.) is followed by pressing the TAP button on the user interface.
VGA: Click on the TAP field

Editing Marker
Markers can be edited as follows:
- Move (VGA only):
  Click on the Marker and drag.
- Delete:
  VGA: Hold the VS's CLEAR button and click on the marker to be deleted.
  VS: Navigate to a Marker. Hold CLEAR and press TAP
- Marker Menu:
  Utility Menu → Marker;
  VGA: VS's SHIFT + click MARKER

2. Arrangement Table
In the menu "Region Arrange" Regions are created from Marker positions and arranged in table form.

> Track → Region → Region Arrange

2.1. Creating Regions:
F1 (CREATE) attaches a new entry column in which the Start and End Markers for a region are to be determined. Positioning at the "Start" (default setting = Marker 0) and the "End" with the cursor buttons. Markers are selected with the dial. The regions created in this way are shown in the playlist for monitoring purposes. To create the following Regions until completion proceed in analog. For corrections, columns can be entered (F4) or deleted (F3 or F2 delete all).

2.2. Specifying the Target Position
F5 (NEXT) calls up the entry page of the target position. The desired values can be entered in the corresponding Timecode or Measure field (the last three fields). The functions →Go To (Positioning) and →Get Now (Time transfer) known from the relative editing operations are also available here.

Tip
Entering the TO Position directly into the →Time Display can be achieved by clicking or with the cursor button ⇕ . The Target Position can also be given here in the form of a Marker. Using F2 (Get Now) the current Time Position will be transferred into the TO field.

2.3. Copying Regions
F4 (OK) copies the Regions created in the list in one step to the chosen Target Time. Further copying to other Target Positions is also possible.

Hint
Because the "Region Arrange" created in the list is saved together with the project data, copying can be repeated at anytime.

Region Compression/Expansion
A region's time can be extended or compressed using Compression/Expansion and its pitch can be changed. This procedure is also called "Time Stretching".

Application:
- Changing the tempo of Drumloops
- Shortening of word contributions
- Changing the pitch

> Track → Region Comp/Exp.

In compliance with →Region Edit the Region that is to be changed is firstly configured using IN and OUT. Depending on the position of TO in relation to OUT, the operation will either shorten or lengthen the Region.

Pitch:
Using "Variable" the pitch changes proportionally to the time changes. The setting "Fixed" maintains the original pitch.

Region Copy

Amp:
In case distortion appears, the volume can be decreased using this parameter.

Changing the Tempo of a Drumloop
The known tempo of a Drumloop is to be entered in the →Tempo Map (SHIFT + Tap). The Drumloop is then to be trimmed exactly (generally one bar) and moved to the timing position 001-01 (→Phrase Edit). With the new tempo navigation at 002-01, this value is then adopted for TO. Region Compression/Expansion can then be carried out.

Region Copy

R. allows the one time or multiple copying of a created Region out of one or more tracks to a desired time position. The original remains untouched in this case (→Region Edit).

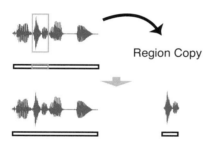

Fig.: Copying a Region

Conditionally, these virtual copies require no additional memory space (→Non-Destructive Editing). A Region can concern one single track or numerous tracks. Region Copy can be carried out either graphically with the mouse, numerically in a table or graphically/numerically using convenient the →Wheel/Button Editing.

Application
- Classic copying of passages to other time positions (Using Choir for further Refrains).
- Drumloops,
- Creating alternative Arrangements etc.

Tip
Copy the Songs several times consecutively with →Region Arrange for:

- Editing variations with extensive Edit Operations
- →Automix-Variations.

For multiple copies of the same passage it is better to use →Phrase Copy. A passage is configured here permanently and doesn't have to be redefined for each editing.
Using the operation →Region Arrange, Regions can be arranged over all Tracks with the help of Markers.

Graphical Copying with the Mouse
Copying with the mouse in the VGA is the popular method of editing.

• Activate Region Edit
In the VS Phrase Edit and Region Edit are alternatively available →Phrase / Region Selection). Region Edit is the default setting and is shown with the red REGION button on the right over the →Measure Bar or:

Right click in the Playlist ➔ Region
VS-2480: PHRASE/REGION button ➔ red

• Creating a Region
Drawing a frame over the desired time and track areas creates a Region. This is additionally identified in the measure bar with the IN/OUT icons (VGA: ⌐ ¬, LCD: ▪ ▪). For corrections, the IN and OUT icons can be moved with the mouse. Clicking in an area outside of the Region cancels it out.

• Region Capture
For this the mouse cursor is positioned over the Region whereby the Icon changes to ✋. By clicking and holding (!), the mouse Icon changes to ✊. Additional pressing and holding of the SHIFT button on the VS creates a "+" in the Icon (✊).

• Region Copy
Drag the Region (frame) with the mouse to the desired position. If the target time lies outside of the currently displayed playlist, →Scrolling over the page can find it. Firstly let go of the Mouse button and then SHIFT on the VS. The copy will then be carried out.

Region Copy

• Support Tools

To →Zoom the tracks in the playlist press SHIFT + Cursor button or click on the "+" or "-" symbol in the scroll bar (→Scrolling). A waveform display can be called up by right clicking in the playlist → WAVE (VS-2480: WAVE button (→Waveform).

The TO time position can be constrained in a magnetic →Grid. In the TO field over the playlist, the exact time position is displayed. During positioning the special cursor Icons give information about the FROM and OUT times (→Cursor).

Graphical Copying via Time/Value Dial

Graphically copying using the Time/Value Dial combines the advantages of mouse editing and numerical entry: clarity and precise positioning (→Wheel/Button Editing (p. 243)).

• Activate Region Edit

The method of operation called "Region Edit" is a default method. In the VGA, the control panel REGION blinks red.

See graphical copying with the mouse.

• Creating a Region ▱ ▱ ▪ ▪

First navigate to the desired Region start time, then press the IN button*. Proceed as an analog function for the region End and press OUT*. The desired track is to be selected and then activated with YES (→Region Edit).

• Deleting TO in the Edit Memory:

CLEAR + SHIFT + TO* deletes the former TO value (→Edit Point Buttons (p. 60)).

  * *VS-2480: Without SHIFT*

• Call up the Copy operation

For the Wheel/Button Wheel/Button Editing do not call up "Copy" via the Track Menu: Right click in the playlist → COPY
VS-2480: COPY button

• Position the Ghost-Region

Time:

Time positioning of the copy can be carried out comfortably with the dial. In →Time Display for this function the cursor (here a red line) can simply be set via ◂▸ to the desired time measurement (hours to subframes). Furthermore Bars, beats and also Markers/Locators and all →Positioning methods can be used. Because in this case the graphic copy isn't always immediately visible, the dial can be turned respectively one grid position forwards or backwards.

Choosing a Track:

The cursor buttons ⇕ move the graphic copy to the desired track.

• Copying

The Ghost Copy becomes a real copy of the original Region by pressing YES. NO can be used to cancel the Copy operation. For additional numerical monitoring, →Edit Message can be used in the LCD.

Numerical Copying

Numerical Copying (→Region Edit, p. 190) can be carried out with the highest accuracy.

Track → Region Copy

The Region is determined by entering IN and OUT time positions. Generally the IN time is adopted for →FROM. TO determines the target position. The Source/Target assignment takes place using F1 (Sel Trk) or directly using the VS buttons (→Track Select, not VS-2000).

+Insert

A copy can overwrite previously existing audio data or move it up to the right in order to make space for the copy with an activated "+Insert" (→Region Insert).

Multiple Copies

Without demanding additional memory storage space, multiple copies can be created with an appropriate entry in the "Times" field.

Hint:

It is much better to create rhythmically exact Drumloops etc. using the →Phrase Copy operation because by using "Quantize", the respective TO positions will be quantized at the exact measure positions.

Region Cut

Using "Cut", Audio Data within a specified Region (→Region Edit) can be deleted. All auto data to the right of this Region then move back the respective amount of time.

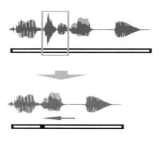

Fig.: Region Cut

This operation is comparable with cutting a passage out of an audio tape and then joining the two tapes together. The following audio signals move to the left and change their position in contrast to →Region Erase. When →Region Mode (default setting) is activated, the regions to be erased can either be deleted by drawing a frame with the mouse or with the →Edit Points IN and OUT, subsequently confirmed with YES. After right clicking in the playlist of the flip menu, "Cut" is to be selected whereby the operation will be carried out. The numerical editing in the Region Cut menu allows a good overview:

Track → Region Cut

Here, when the parameters are activated, →All V-Tracks including inactive V-Tracks are included in the editing. Erased Regions can be recalled at any time using →Undo.

Application
For arranging (e.g erasing a verse) and moving subsequent passages back.

Region Edit

Any temporary passage which is especially created for one single editing procedure is called a Region. IN and OUT time points define the passage and it can stretch over one single track or multiple tracks. Editing only occurs within these boundaries. In contrast to a Phrase (→Phrase Edit), a Region doesn't exist permanently. Subsequent edit operations write over the current Region.

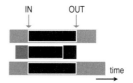

Fig.: Region Edit

In the VS the following operations in →Region Mode are available:

- Copying and Moving
  →Region Copy (p. 186)
  →Region Move (p. 191)
- Deleting and Removing*
  →Region Erase (p. 190)
  →Region Cut (p. 188)
- Inserting a blank passage*
  →Region Insert (p. 191)
- Importing and Exporting of tracks
  →Track Import (p. 228)
  →Track Export (p. 228)
- Arranging songs using Marker
  →Region Arrange (p. 184)
- Expanding and Compressing Time
  →Region Compression/Expansion (p. 185)

* Editing all V-Tracks of one Track is possible (→All V-Tracks)

Regions are also used in →Automix Edit and in the →Phrase Sequence of the VS-2480 in order to carry out corresponding operations within these boundaries. Editing of a Region occurs in the same manner as the similarly dealt with →Phrase Edit by hard disk recorders as →Non-Destructive Editing and is able to be cancelled at any time using →Undo. Region Editing can be practically carried out using →Numerical Editing →Mouse Editing and →Wheel/Button Editing. In stereo coupled track channels (→Link), left and right tracks are edited together. It is necessary to pay attention in particular

Region Edit

to whether the →Record Modes of operations with →Master Tracks (→CDR) match.

Application:
Edit operations that are only carried out once such as copying, moving passages for rhythmical correction, time stretching etc.

Tip
Because of the fixed passages, multiple copying and moving for Arrangement purposes can be better realized with →Phrase Edit.

Region operations are carried out in accordance with the following steps although graphical editing automatically carries out some steps:
1. Activate the "REGION" Edit Mode
2. Region create
3. If necessary, save the TO time position
4. Carry out the Operation

Edit Points
The time positions, necessary for creating a Region and the target position are specified in the VS using the standardized edit points IN, OUT, TO and FROM. These are entered directly either numerically or with the Buttons/Dial editing. Purely graphic copying with the mouse automatically generates Edit Points, which can be read above the playlist from the created Icons.

Fig.: Edit Points and Region

The diagram shows the display of a Region on the VGA monitor (without →FROM ▽). In the LCD of the VS the Edit Icons are shown as squares.

IN	OUT	FROM	TO
Region Start	Region End	Anchor point	Target time

→Edit Points can be generated, moved, deleted, and corrected. In this way, moving an OUT or IN point with the mouse changes a created Region easily.

Edit Memory
The Edit Memory can be used for simplified entry of Edit Points directly using the control panel and without having to call up the respective Region Menus (not necessary for graphical editing with the mouse). To do this, position the project at the corresponding time and save the relative Edit point with the Buttons for IN, OUT, FROM and TO. This method can be used advantageously for entry corrections. Alternatively this can be done in the VGA by clicking on the respective Icons in the Edit Control list above the →Measure Bar.

VS-2400
Hold SHIFT and press the labeled Locator buttons (4,5, 9 and 0).

Entry Methods
To create a Region as a condition for the various region edit operations, graphic →Mouse Editing, →Wheel/Button Editing and →Numerical Editing in the respective menus can be used. A mixed application of this method is practicable and sensible. Detailed descriptions are to be found under the keywords of the special region operations.

Creating and Correcting of Regions via Mouse
Not only in the VGA monitor but also in the LCD, a Region is generated by drawing a frame in the playlist with the mouse. This is determined horizontally through the chosen time area and vertically through the corresponding tracks. It is displayed as a darkened area.

Correction:
In the VGA monitor, the IN and OUT Icons can be moved with the mouse and the Regions can be time corrected in this way.

Deleting:
Holding CLEAR and clicking on an Edit point deletes it and therefore the entire Region.

Region Erase

Adding Tracks

Further Tracks can be added to the Region using SHIFT and clicking in these Tracks, whilst clicking outside of the Region deletes them.

Creating and Correcting a Region with Dial/Button Editing

Button/Dial editing offers higher accuracy in addition to good clarity. At the desired start time position the IN point is entered using the edit memory (see above) and at the Region End, the corresponding END point. An entry correction can be carried out in analog form. The existing values are automatically overwritten (When necessary pay attention to the →Edit Point Sw Type, which changes the behavior of the buttons). The desired Track is selected with Cursor Up/Down. Pressing YES generates the Region. Additional Tracks can be added and removed from the Region using SHIFT + YES. Holding CLEAR and calling up an Edit point deletes its entry in edit memory.

Numerical Creation and Correction of a Region with the Region Menu

Numerical editing offers the highest accuracy and can additionally be carried out musically with a Lead Sheet or Score using the entry of bar numbers.

Track → Region xxx

The Region is specified through the Entry of the IN and OUT time positions. Generally →FROM adopts the IN time. When necessary, TO determines the target position. →Get Now (for adopting the current time in the specified field) and →Go To (for positioning at the time position of the time value from the current field) act as support tools.

The time values in the Edit points fields can be entered with the dial, the mouse or also numerically (→NUMERICS [not on the VS-2000]). Corrections result from a new entry.

Choosing a Track

The Region can include one or more freely chosen Tracks. Choosing takes place in the menu called up via F1 (Sel Trk) or via the corresponding button assignment (→Track Select).

Region Erase

With "Erase", Audio data within the indicated Region can be deleted. This operation is comparable to deleting a tape passage. The time position of the subsequent recordings remains untouched unlike in →Region Cut.
Details: see→Region Edit (p. 188)

Execution

When →Region Mode is activated (default setting) the Region* to be deleted is specified by:

Drawing a frame with the mouse   or
Setting the →Edit Points IN and OUT for the start and the end of the region and selecting the tracks by subsequently pressing YES.

After a right click in the playlist "Erase" can be selected from the flip menu. The operation can then take place.

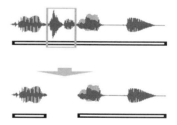

Fig.: Region Erase

A complete overview is allowed by numerical editing in the Region Erase menu:

Track → Region Erase

When the parameter →All V-Tracks (p. 15) is activated, all V-Tracks from a track are included in the editing.

Deleted Regions can be recalled at any time using →Undo.

Application

Deleting passages within a Track whilst keeping subsequent passages.

Region Move

Tip
- Short, unwanted passages at the Start or End of a phrase can be removed more simply using →Phrase Trim (p. 157).
- An entire Phrase can be removed more quickly via →Phrase Delete (p. 149).
- An entire Track is better deleted without entering Edit Points via →Track Exchange.

Region Insert

"Insert" lets a blank passage be inserted at the indicated position (→Region Edit). This operation is comparable with separating an audio tape and inserting an additional blank band - the following audio signals move up to the right and thereby change their positions. When →Region Mode is activated (default setting), the blank passage to be inserted can be specified either by drawing a frame with the mouse or using the →Edit Points IN and OUT and subsequently pressing YES (→Region Edit).

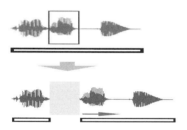

Fig.: Region Insert

After right clicking in the playlist "Insert" is selected from the Flip menu whereby the operation takes place. →Numerical Editing allows a complete overview in the menu:

Track → Region Insert

When the parameter →All V-Tracks is activated, also inactive V-Tracks from a track are also included in editing. The operation can be recalled at any time using →Undo.

Application
- New arrangements
  (e.g making room for a new Verse)
- Moving subsequent passages forward.

Tip
To insert a recorded passage instead of a blank passage, the operation →Region Copy with the " + Insert" parameter activated can be used more quickly.

Region Mode

In the VS two operation modes for editing audio passages are available: →Phrase Edit and →Region Edit. Changing respectively to the other mode can be done in the following ways:

VGA-Display
- With one click in the REGION field (red) the mode changes to PHRASE (green) and back (Playlist →Measure Bar on the right).
- Right clicking in the playlist. In the pop up menu. Phrase or Region can be chosen.
- The mode of operation changes automatically through choosing the corresponding entry in the TRACK menu.

VS-2480:
PHRASE/REGION button (green/red)

Track Menu
In the LCD the mode of operation can be changed using F6. The indicator changes correspondingly Reg→Ph or Ph→Reg.

Region Move

Move is used for positioning a Region to a new time position whereby in contrast to →Region Copy, no copy is made. It takes place in identically the same way as Region Copy. Multiple placements cannot be carried out.

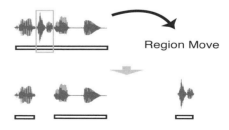

Fig.: Moving a Region (Move)

Region/Phrase selection

Application:
- Move Song Parts for arrangement changes.
- Moving chosen Tracks to improve Timing

Region/Phrase selection
→Phrase / Region Selection (p. 157)

Rehearsal
R. represents a simulated Recording resulting in changes for listening conditions particularly in the →Headphone Mix. This mix is generated for musicians via Aux busses in the Track Mixer. Thus it appears that they could being able to only hear their own instrument during the actual recording or when the project is stopped. So the Utility Parameter →Record Monitor (p. 175) determines the behavior of armed tracks.

Relative Time
During →Synchronization in Slave mode, the incoming Timecode can be displayed in any time offset for reasons of clarity. The synchronization of the Master and Slave doesn't change by doing this.

Utility → Project → Time Display Format

VS-2000: Utility Menu → Display Parameter.

Release
In the →Dynamics, the completion of the Compression/Expansion can be delayed via the Release time. In doing so, among other things, the so called "Pumping" of a compressor can be avoided. This occurs when the processor is unpredictably triggered by unwanted elements of a sound.

Reload
The →Project operation R. allows loading the original version of the current project whereby the current state after the last →Save is reestablished. All recordings and editing operations not yet saved are irretrievably lost in this procedure. R. is therefore used for the general canceling of editing operations in an opened project. →Undo (p. 231) is principally able to lead to the same result. Due to the fact that Undo keeps all previous recordings, it requires additional memory space on the harddisk in contrast to Reload.

Remain
The Remain field in the footer of the VGA shows the remaining hard disk capacity chosen as either time, MB, % or remaining →Events (→Remain Display Type).

Remain Display Type
Numerous parameters are available to determine the remaining hard disk capacity for the recording. The corresponding model is then shown in the Display*:

Time:
Recording time is for one mono track. For a stereo track, halve this time. For four tracks, quarter it etc.

CapaMB/Capa%:
The display in MB or % is less practical and is generally only to be used for a USB backup (VS-2000).

Event:
Independent of audio data, the →Events in the VS are used to determine the duration of tracks and phrases, the →Automix etc.

> * VGA-Display: permanent displayed (→Remain)
> LCD: instead of the date display
> (→Date/Remain Sw)

Remaining Space
→Remain (p. 192)

Removable Disk
When the VS-2480 was first shipped, it possessed a removable disk cartridge instead of the newer model's integrated CD or DVD drive. This allows simple exchanging of projects for devices of this model range. A suitable CD burner is to be attached here via →SCSI.

Reset
Mixer and Utility settings can be recalled to normal values using →Parameter Initialize (p. 142).

Resolution
The resolution (word length, Bit rate, quantization) represents the size of the grid of a digital audio signal, which is vertically aligned for the volume. R. determines the number of level steps and the possible →Dynamic. Higher Bit Rates improve not only the Dynamics but also the signal-to-noise ratio through fewer rounding errors which represent the origin of quantization noise. Through this the amount of data to be edited and saved increases in size, which makes the more effective DSPs and larger hard disk capacity necessary. Aside from the R., the →Sample Rate displays a further quality criterion for digital audio signals. In the VS, the R. is configured through the →Record Mode when creating a new →Project. Here maximum 24-Bits are possible. Internal signal operations in the VS are carried out with 56-Bits.

Restart
To give the hard disk a longer break it is not necessary to completely turn off the VS. Instead the →Shutdown (p. 206) procedure is enough. Using R. the VS can then be quickly rebooted.

SHIFT + PLAY

In the VS-2480, this operation can also be used to continue after ejecting a removable disk.

Return to Locate Sw
For security reasons, set as a default, after a →Scene operation, the VS changes back to the →Locator operation mode. This prevents accidentally calling up Scenes and overwriting current Mixer Settings. To deactivate this protective setting for special applications, the following steps are to be carried out:

Utility → Global1 → Return to Locate Sw = Off

Return to Track Status
In the VS-2000 the functions →Track Status, →Locator and →Scene all share common buttons. After calling up a Locator either this setting "Locator" can be kept or it can be changed automatically back to the "Track Status":

Utility → System1 → "Rtn To Tr Track Status Sw"

Reverb
The algorithm R1 (Presets P036 - 053) can only be used to odd numbered →Effects processors and blocks the second processor for certain effect patches. This restriction is not applied to R2 (Presets P054 - =73).

Reverberation
Hall (Reverb) represents the entirety of countless statistically distributed reflections for reasons of room reflection, which chronologically diffuse and melt into one another and which gradually fade out.

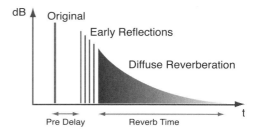

Fig.: Early Reflections and Reverberations

Setting parameters
The Reverb algorithm (1) is provided with a 3 band →Parametric Equalizer in order to filter the input signal to be reverberated. In this way specific frequencies can be cut out of the effect processing (e.g. in an entire drum track the Bassdrum and Cymbals/HH shouldn't be processed with reverb effects). The EQ in the Algorithm "Reverb2" is located after the reverb block whereby the effect signal filters itself. Because an Equalizer isn't available in the →FX Return Channel, this represents a practicable editing possibility.

Rhythm computer

Reverb

Both the Presets of the Reverb2 Algorithm and the →Plug-in Halls offer a choice of so called Hall types (Room, Hall, Plate). In this way, the Hall characteristics such as room geometry, type of reflection surface, fading behavior of high and low frequency rates or the corresponding simulation in the form of default macros can already be simulated in advance. In the Preset R2 however, the corresponding Early Reflections are included in the macro and cannot be separately set. While "Room Size" determines the size of the virtual hall room, "Time" changes the fading time of the diffusing reverberations ("Hall length"). "Density" affects the thickness of the Reverberation, which is responsible for the number of diffused reflections. To separate the reverberation from the original signal the reverb can be delayed using "PreDly". This makes using bigger reverb rooms possible without "smearing" the original. Values of up to 100 ms can be used here. Naturally sounding rooms are characterized by dampened high bands and a faster fading out of higher frequencies. With its brilliance however, a modern vocal reverberation supports the voice and the melody lead instruments being in the foreground. In the VS different fading behaviors of high and low frequencies of the effect rate can be simulated with the LF and HF Damp Parameter. Cutting the level of a Band using GAIN here seemingly shortens the reverb length for a frequency range while the other respective areas maintain their fading time, which was predetermined with "Time". For experimental applications, absolutely controversial settings such as Size = 40m and Time = 0,1 s can be used with an input delay for the reverb of 200 ms using "PreDly".

Early Reflections

The parameter "ERLvl" controls the volume of Early Reflections (0 = no ER), whilst "Diffusi" specifies the number of the first reflections (choose smaller values for vocals). The pre-delay has an equal affect on the ER and the diffuse reverberations.

Rhythm computer

→Rhythm Track
VS-24xx: →Metronome

Rhythm Track

Only the VS-2000 provides an internal fully programmable rhythm computer with selectable drum sounds. Self created Patterns or default →Presets can be used to create one's own Rhythm Arrangements (Songs, Rhythm Arranges). Alternately, predetermined ones can be used.

1. Activating

Using the R. excludes audio tracks 17/18 from playback. Either the Rhythm Track or the Audio Tracks are then assigned to the Stereo Mixer channel 17/18:

VGA: Rhythm Track → Rhythm Track → Track 17/18 Assign = Rhythm

VS-2000: RHYTHM TRACK → Track 17/18 Assign = Rhythm

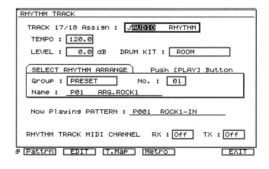

Fig.: Rhythm Track (Main Page)

Hint

The Rhythm Track is also activated for play by using the TRACK/STATUS buttons.

2. Arranges (Songs)

2.1. Selection and Playback

The 49 Preset or self created "Rhythm Arranges" (Rhythm Arrangements, Songs) can be selected in the main page of the Rhythm Track. To do this, choose "Group":

Rhythm Track

- Preset: 49 internal Songs of differing stylistics, which cannot be overwritten.
- User: 10 Memory locations for self created Rhythm Arranges, which can be used comprehensively over all Projects.
- Project: 10 memory locations for self created Rhythm Arranges only for the current project.

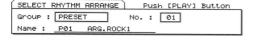

Fig.: Selecting a Rhythm Arrange

By starting a Project, the chosen Rhythm Arrange will be played with the Settings made in the Tempo Map and the currently played Pattern will be shown. All parameters of the Mixer channel 17/18 accessed via →CH EDIT can be used for the sound editing of the Rhythm Track (EQ, Dynamics, Effects etc.).

2.2. Tempo and Type of Beat

Rhythm arrange accesses the programmed Tempos and beat types in the →Tempo Map, which are identical to the →Metronome. The Tempo Map can either be called up using F3 or with SHIFT + TAB.

2.3. Personal Arranges (Songs)

Personal Arranges can be put together from Preset or User patterns.

Hint:
All newly created or edited Arranges must be separately saved as User or Project Arranges. Selecting another Arrange or if the VS is →Switch off, unsaved arranges are lost.

The Arrange Edit page can be called up using F2 (Edit). The current Rhythm Arrange can only be deleted after multiple "Cut" operations or by selecting an empty User Arrange. Using "Group" (Preset, User Project) and "No.", the first Pattern can be chosen and listened to with F1 (Preview). The name of the Pattern, its type of beat and original length will also be displayed. Generally the first Pattern will be set at Bar 1.

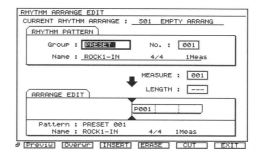

Fig.: Edit page in Rhythm Arrange

Positioning in an Arrange:
Using FF and REW, the →Timeline can be positioned by measures. When the measure field is selected, the dial can also be used for choosing measures. Simultaneously holding SHIFT enables positioning in a 10 bar grid.

MEASURE:
This field shows the current measure position and serves to position it in the Rhythm Arrange (See above). It is identical to the beat indicator →Time Display.

Overwrite
The chosen Pattern can be used via F2 at the Measure Position. A previous Pattern will be overwritten by doing this, although the original pattern length will be maintained.

Insert
F3 places the new pattern at the Measure Position and moves the following patterns correspondingly.

LENGTH
L. specifies the length of the current Pattern. Larger values than the original lengths can also be chosen. The pattern will then be correspondingly repeated.

Cut and Erase
To delete a Pattern at the Measure position, →Erase creates a blank section in the Rhythm Arrange whilst maintaining the subsequent pattern positions. By contrast, the →Cut opera-

Rhythm Track

tion subsequently moves the whole pattern up after removal which is why it is suited to shortening the entire Rhythm Arrange.

Assign
A. helps for listening to and trying out arrangements by allocating patterns to buttons 1 - 17/18.

2.4. Editing a Rhythm Arrange

Rh. Arranges can be copied and deleted. This is convenient for copying Project Rhythm Arranges to User memory locations for universal applications, to create security copies etc.

- Call up the Rhythm Track
- Assign COPY and DELETE to F1 and F2 via PAGE (if necessary)
- In the Group field, select the designated Arrange (Copy: Source field)
- Copy: Choose a memory location in the Destination area via "GROUP"
- F5 (OK) carries out the Operation.

2.5. Internal and External Saving

Rhythm arranges can be saved internally to the VS hard disk or also burned to a CD in order to be able to use them universally. CD burning takes place in the main page of Rhythm Tracks, accomplishing an analog operation to →Project Backup (p. 166). In the Edit page of Rhythm Arrange, created or changed arranges (Songs) can be saved either as a project comprehensive User or Project related Project pattern. →Name Entry can then take place.

3. Pattern

Aside from the 295 default Preset Patterns, 999 additional User and a further 999 additional project patterns can be created. As recording methods "realtime", "Step" and "Micro Edit" are available here. Generally the STATUS buttons are used to conveniently enter the drum instruments. On the VS panel, these are labelled correspondingly.

3.1. Choosing a Pattern

To create a Pattern, firstly call up the Setup Menu using F1:

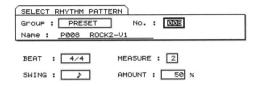

Fig.: Setup page for Pattern Entry

Choose an empty User or Project Pattern and carry out the corresponding settings for the number of beats, type of beats and the Swing Factor. Principally as a basis, a preset patter can also be used and single Instruments can simply be deleted.

3.2. Recording in realtime

F3 calls up the realtime page and the input tempo, quantization and drum set can additionally be chosen here. In this mode of operation, the STATUS buttons serve the entering of Instruments. Pressing REC followed by PLAY the recording starts after a one count in beat. Because this is implemented as a Loop, the respective Drum instruments can be inputted numerous times.

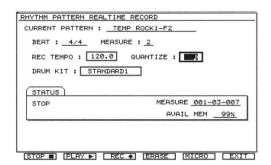

Fig.: Inputting Patterns in realtime

Without interrupting the recording, F4 (Erase) allows drum notes to be deleted at the respective count time and to return to the recording.

3.3. Micro Edit

Using F5 "Micro Edit" is accessed. Micro Edit lists drum notes in tabled form, which gives a clear overview for corrections. By pressing

YES and positioning the Cursor on an entry, all parameters can be changed and also deleted. Micro Edit can also be accessed from Step Edit.

3.4. Step Edit

Fig.: Step Entry for Patterns

Step Edit is particularly suited to difficult rhythms because of its graphical entry of drum notes using two different methods. F4 calls up the Record Step menu from the Pattern Setup page. Firstly choose the smallest rhythmical resolution using "Step". Along with the number of columns, this specifies the step width for positioning and is generally measured in 1/16. The current entry method is displayed on the top right under the STEP field. Toggling between both modes takes place with F1 (Rec-Mod). When necessary, press Page beforehand.

3.4.1. Button Mode

After positioning the Cursor in the CURSOR field, the dial is used for positioning with the rhythmical step width specified above or with F3 and F4 in measures. Pressing one of the STATUS buttons then generates the desired note. Repeated pressing deletes it again.

3.4.2. Dial Mode

The Dial mode is a purely graphical entry method whose distinctive feature is a continuous specification of note velocity (volume). After being activated with F1, the cursor is positioned over the corresponding time unit (row) and the desired instrument (columns). The note velocity is controlled here with the dial, which is graphically represented with a circle's diameter.

3.5. Copying and deleting Patterns

Patterns can be copied from Preset, User and Project groups optionally into one another and then further edited. In this way project oriented patterns are able to be used in other projects. Only presets can't be chosen as the Destination.

3.6. SMF Import for Rhythm Patterns

From a CD-R (ISO 9660 format), eight bars in a User or Project pattern are imported from the drum track of an SMF (Midi Channel 10):

Rhythm Track → F1 (Pattrn) → Page → F3 (SMF Imp)

The desired SMF is selected from the displayed List using F5 whereby the SMF Import page appears:

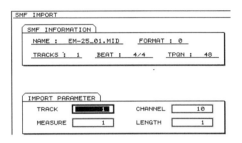

Fig.: Specifying Import parameters for SMF

For the max. eight bars (Length) that can be imported, the Start bar can be stipulated with "Measure". F5 starts the operation and the SMF is saved and stored temporarily. Here it can already be pre-listened (press page). It can then be saved to a free Pattern.

F5 (Save) → Group x and No.xx →
F1 (Name) F5 OK → YES

Rotary

This →Effect, mainly used for organs to create the typical Leslie sounds, simulates the sound of two rotating loudspeakers.

Rotary	
Effect Patch:	P 015
Connection:	Insert
Application:	Organ

Routing

Additionally, a →Noise Suppressor to suppress noise and an Overdrive for a distorted sound are available.

Routing

R. describes not only the procedure of virtually wiring from sources to targets in the VS but also the results of this. Aside from the methods shown under →EZ Routing, the →Patchbay can be called up in the LCD as follows:
- Select the →Input Mixer
- Choose a channel via SELECT
- F2 (P.Bay) selects the Patchbay
- Activate an input jack pair with the cursor buttons and the dial.

Hint
The connection between inputs and the Input Mixer can only be made in pairs.

Routing View

The R. menu page allows routing the inputs to the →Input Mixer and assigning Recording tracks to the corresponding outputs of the Input Mixer (→EZ Routing).

RPC-1

The PCI audio card for Mac and PC allows eight digital audio signals to be transferred via the →R-Bus from the VS to a computer (not in the VS-2000) and back (bidirectional). According to the R-Bus format, the data format is max. 24 bit/96 KHz (→Word Length, →Sample Rate). The transfer of →MIDI data and the →Word Clock occurs parallel to the audio data. The audio card is only still available in the second hand market. The new Audio interface "→V-Fire" made by the company Presonus transfers VS Tracks via Firewire.

RSS

"Roland Sound Space" positions a signal over the stereo field of the two monitor speakers. It creates an intense spatiality in the →Effects, which take advantage of RSS.

RSS PAN

The →Effect R., which enables positioning a sound source in a circle around the listener using two speakers or headphones. It is only available on the VS-2000/2400 models. Allocation occurs in the UTILITY menu or more simply, directly in the corresponding Mixer Channels.

CH Edit → Click on RSS Pan
Menu → Utility RSS Pan Setup

A board for RSS Pan must firstly be assigned in the "Use Effect Board" field, which then enables RSS PAN for six* independent Mixer Channels. The board is then no longer available for other effect applications. In the table, the desired Mixer Channels for RSS can be selected. Additionally an algorithm optimal for headphone use can be activated. The respective Pan knobs of the selected channels change to the RSS Pan circle and position the signal via Mouse or Dial entry correspondingly. Although it doesn't create an ideal circular motion around the listener, the effect still generates an exiting amplitude far above the normal stereo basis. RSS Pan is also audible in Solo Mode.

* *VS-2000: four*

RSS Pan Grade Mode

In the →Automix's SETUP this special →RSS PAN parameter restricts the signal positioning of a movement, which was created by a gradation. In front of the listener (Setting: clockwise) or behind the listener (counter clockwise).

RSS Pan Setup
→RSS PAN

S

S/PDIF
The Sony/Phillips Digital Interface represents in its two-channel Interface a popular digital consumer audio format. It is based as far as possible on the professional →AES/EBU and is in this way self clocking. The necessary →Clock is contained in the data stream. A copy protection flag (→SCMS) prevents digitally copying second generation dubbing. Two physical transfer forms are applied for the S/PDIF signals, coaxial and optical. In the VS these are firstly to select for establishing a connection (→Recording of digital Signals).

1. Coaxial (electronic)
The electrical transfer mode, also described as IEC-958 TYP II occurs with a signal amplitude of max. 0,5V and should take place via a special coaxial cable (→Digital Cable). The connection is →Unbalanced and is applied via the RCA connection (Cinch).

2. Optical
The optical transmission of digital audio is used by light pipes. A special LED in the Sender creates the light impulses that correspond with →Binary Numbers. The receiver's photo cell takes over the conversion back to electrical impulses. The optical connection was patented by Toshiba and therefore called TosLink, which is also used for the →ADAT format.

Safety in Operation
Stand Alone hard disk recorders in contrast to PC supported audio sequencers have a higher operation security due to the special audio hardware and defined system software. Nonetheless, particularly the →Hard Disk is subject to →Fragmentations and attrition typical of computers. To maintain a high level of operational security it is recommended to →Format the hard disk in connection with a →Surface Scan in regular intervals and at least once a year after undergoing the necessary →Backup procedure. It is likewise recommended to carry out a →Drive Check, which checks the table of contents data of a →Partition (TOC). To →Switch off always use the →Shutdown procedure. Through this the recordings, editing and Settings last carried out are saved, the write/read heads of the hard disk are set at a safe standby position and the hard disk is shut down properly. Simply turning off with the main power switch can lead to losing data and/or damaging the hard disk. For transporting, leave the hard disk to spin down for two minutes.

Sample Rate
By sampling the analog signals in the →A/D Converter, the S. establishes the number of samples in one second and therefore directly determines the maximum frequency of the digitalized audio signals. A Sample Rate of 44,1 KHz corresponds to 44.100 samples per second (see Oversampling). According to the Nyquist theorem, the highest frequency of the audio signal that is to be reproduced requires a doubled sample frequency whereby for the S. of 44,1 KHz a maximum threshold frequency of approx. 22 KHz results. Each of the single Samples provides information about the voltage at the respective time of extraction. The measuring accuracy is here determined by the →Word Length (Bit rate). Digital signals that are to be recorded must exhibit the same S. as the current VS-Project or the digital transfer cannot be carried out properly. For this, the →Master Clock is to be correspondingly set. Except for the VS-2000, which only records audio material with 44,1 KHz, the Sample Rates of 96, 64, 48, 44,1 and 32 KHz are available for the VS-24xx models. Using the →Vary Pitch function, the Sample Rates can be varied slightly (not in the VS-2000).

Hint:
Digital dubbing of →Audio CDs requires a Sample Rate of 44,1 KHz.

Sampler
With →Phrase Pad (p. 153) and →Phrase Sequence, the VS-2480 possesses a Sampler function.

Save

Save   ▪▪▪

After recording and other important operations, always save the current project regularly. (→Project Store (p. 169)).

> SHIFT + ZERO
> (STORE)

Scene   ▪▪

Saving and recalling Settings (→Total Recall) represents one of the greatest advantages of digital mixers and hard disk recorders in comparison to their analog counterparts. In the VS, 100 memory locations are available as so called Scenes for saving →Mixer-, →Routing-, →Effect and →Track Status/→V-Track settings. When setting up a new project, scenes can also be transferred with →Copy Mixer Parameter. A →Scene Import/Export is also possible via Midi. The VGA displays the current Scene with its number and name.

Hint

- The monitor level (knob) is not saved
- Single Mixer channels can be protected against being called up in a Scene.

Application

- Different Routings and listening settings for the recording.
- Changing Mixer settings; →Effects; V-Tracks for Mixing and Mastering

  * *Using Scenes in →Automix
  →6.2. Scene To Snapshot (p. 24)*

In contrast to →Snapshots, which have a limited number of Parameters to store, Scenes can only be saved and called up when stopped.

Saving and Calling up Scenes

In total in the VS, up to 100 Scenes are available and these are organized in banks. The models VS-24xx provide a comfortable and direct access to the 10 scenes of the current bank via their numerical buttons. The VS-2000 however, uses the STATUS buttons for this as a double function.

1. VS button

A scene can be comfortably and simply saved and called up again in the current bank using the corresponding VS button. After saving or calling up, the VS automatically changes back to the Locator Mode. Doing this avoids writing over Settings by accidentally calling up Scenes (→Return to Locate Sw).

- The Scene Mode must be activated (SCENE button blinks).
- Pushing a free, not lit button 0-9 saves Settings to this Scene Memory.
- Pushing an occupied (flashing) button 0-9 calls up the Settings of a saved Scene.
- Holding CLEAR and pushing an occupied (lighting) Scene deletes it.

Scene Bank:

> SHIFT + Scene

The blinking buttons refer to the current Scene Bank. Activating a blinking button changes it to that particular Scene Bank.

VS-2000

The VS-2000 provides 96 Scenes in 6 Banks of 16 Scenes. Firstly select the Scene operation mode.

> SHIFT + LOCATOR/SCENE.

Herewith the STATUS button acts as the Scene buttons 1-16 for the current Bank.

Scene Bank:

> SHIFT + LOCATOR/SCENE → 2 s hold.
> Select Bank No. → ENTER.

2. VGA

By clicking in ◄ in the →Measure Bar on the right, the PlayList control field is opened. The Scene field can then be called up via "Scene":

- Clicking on a free (not lit) Scene Number saves the Settings in this Scene Memory.

- Clicking on an occupied, lighting Sc.nr calls up the Settings of this Scene.
- Holding CLEAR on the VS and clicking on a blinking Scene number deletes it.

Fig.: Scene in the playlist control field.

Scene Bank:
By clicking on a tab in the Bank row, one of the ten Banks is called up.

3. Editing Scenes

The special Scene Utility page offers an additional way of administration and name entry in the form of a table.

Menu: Utility → Scene
VGA: Click on [SCN]

Fig.: Scene Table

In the table, Scenes are listed with their names and date/time of creation. Firstly select the particular row, the using F3 (Store) creates a new Scene with the current mixer settings. F2 (Clear) erases a selected Scene. Using F1 (Name), a →Name Entry is also possible.

Excluding Mixer Channels:
F5 (CH Sel) calls up the selection window for Mixer Channels that are to be included in a scene recall. Marked channels will be regularly edited through setting changes via Scenes. An empty box in front of a channel number means, that the current channel settings will be maintained even though a Scene change is carried out - The Mixer Channel is then protected against a Scene change. This function can be used for pointed changing of chosen Mixer Channels by calling up Scene.

4. Safety

Because the current Settings are overwritten by a Scene change (when this hasn't been saved as a Scene), precautionary measures can be taken against accidentally activating a VS button:

4.1. Return to Locator Mode
One of the default Settings is that the VS button automatically changes back to the Locator Mode after calling up a Scene or Saving. Doing this avoids deleting currant Settings by accidentally calling up Scenes. This setting should be kept for →Return to Locate Sw (p. 193).

4.2. Save Mode
The "Save Mode" for calling up a Scene represents a somewhat time-consuming but safe operation mode. This mode is only provided for operating on the VS and not in the VGA.

Utility → Global*1 → "Locator/Scene Type" → Save

Firstly activate the Scene mode by pressing the SCENE/BANK button, the EXIT/NO LED button will blink. The VS then demands the entry of a two digit Scene Number. This is to be confirmed with Enter whereby the Scene change occurs. Subsequently, the VS deactivates the Scene Mode again. With an Entry in an unoccupied memory location, saving takes place following the direction.

* *VS-2000: System instead of Global. Confirmation only via YES*

Scene Import/Export

Scene Import/Export
→Scenes of the current project can be sent via →MIDI Bulk Dump to a Midi sequencer or another VS of the same model. From there they can also be transferred back again. This still represents the only possibility to transfer a Scene into another project or to use in other VSs.

> * When creating a new project, Scenes can be transferred via →Copy Mixer Parameter.

Scene P.C Rx Sw
To call up →Scenes via Midi →Program Changes, the Parameter S. must be set to ON (→MIDI Mixer Settings (p. 124)).

> Utility → MIDI Parameter →
> «SCENE P.C. Rx Sw» = on

SCMS
The Copy Prohibit Bit (Serial Copy Management System) is contained in the digital data stream of →S/PDIF signals and prohibits second generation digital copying (only one digital copy is possible). In the VS one's own recordings can also be protected. The parameter "Digital Copy Protect" achieves this (VS-2000: Digital Parameter):

> Utility → Project Parameter* → Digital Copy Protect

Screen Saver
With the exception of the VS-2480, animated Screen Savers are available for the VGA monitor and the LCD. On the LCD the text contents of the COMMENT field can be displayed additionally with a delay setting of more than one minute (→Name Entry).

> Utility → Global 1→ Screen Saver

Scrolling
Vertically and horizontally moving the screen contents to find Objects, which are not currently shown in the display window, is called S. The →Playlist can easy be scrolled by the VGA's scroll bars and buttons.

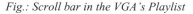

Fig.: Scroll bar in the VGA's Playlist

The cursor buttons ⇅ are also available for vertical alignment (zoomed Track Display).
To move horizontally (on the Time axis) in addition to the scroll and →Positioning tools, the position bar can also be used in the VGA. It is to be found directly under the →Time Display:

Fig.: Positions bar with handle

By clicking and dragging the handle with the mouse, the →Position Line is set at the desired time position.

Scrub
S. plays back a small area of the Track audio at the →Position Line repeatedly. This represents a small Audio Loop. In this way due to the characteristic Loop Noise determined by the audio material and together with "Positioning with the Dial", Audio events are easy to find.

Application
Accurately finding Audio Passages for →Editing and setting →Locators and →Markers.
1. Firstly select the corresponding Track in the playlist by clicking on it (→Track Select (p. 229)).
2. The Time line has then be positioned shortly before the Passage that was searched for (→Positioning).
3. Activate Scrub:

VGA:	Click in the SCRUB-Field
Mouse:	Right click in the playlist
VS:	SCRUB button

4. For Events whose times lie behind the Time line, activate TO (See fig. below), otherwise press FROM.
5. By setting the Cursor (here a red line) in the →Time Display on Frames or Sub-

frames and by approaching the Audio Event gradually with the Dial, the characteristic Scrub noise changes so that a Passage can be found acoustically.

6. Together with the →Waveform display and a corresponding enlargement, both acoustic and graphic monitoring is available.

Pay attention to the Position of the Scrub Loops behind or in front of the Time line, which are specified by the →Preview's To or From. To search for Audio Events on the right of the Time line (later), choose TO. Events before the time line (earlier) can in contrast be found with the Function Preview From. This is made clear in the following diagram:

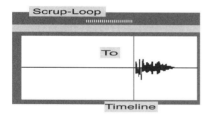

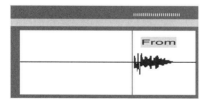

Fig.: Scrub Loop with TO and FROM

The length of time of a Scrub Loop is changed using the →Scrub Length parameter in a range of 25-100 ms.

Hint:
When Scrub is activated, no Edit Operations can take place:

"Cannot Edit! Project is Playing now"

Scrub Length
The length of the short Audio Loop that is activated with the →Scrub function can be changed according to its application (25-100 ms). For microscopic positioning in the Audio Material, choose extremely short times. Generally, the default setting of 45 ms can be used universally.

Utility → Play/Rec2 → Scrub Length

SCSI
"Small Computer System Interface" describes a standardized bidirectional interface, which is being replaced more and more by the IDE interface with reference to the hard drive connection. Among the VS models, only the VS-2480CD provides an SCSI connection for the connection of an additional hard drive. The VS-2480 (removable disk model) uses the SCSI for the connection of the →CD Writer. Pay attention to terminating the last device in the SCSI chain, which generally is achieved with a micro switch in an active Terminator within the external device.

SCSI Self ID
Within an →SCSI connection, all devices require their own addresses, or a so called ID. For the VS-2480, this setting can be changed as follows:

Utility → System Parameter2 → "SCSI Self ID"

The factory default setting for S. is set to "7". Changes become active after restarting the VS.

Section
In Automix, →Selecting an Automix Parameter (p. 22) is necessary for choose other parameters than volume and panorama.

Sector
A sector of a →Hard Disk represents the smallest memory unit. Sectors and Tracks are created while →Formatting the drive (Low Level). Many Sectors form a →Cluster, which defines a memory area of items that belong together.

Select
→Track Select (p. 229), Phrase Select (p. 151), →SelPhr (p. 204)

Select Project
→Project Select (p. 169)

SelPhr
(Select Phrase) Within the →Phrase Edit operation menus, Phrases can be selected for editing from the respective Track using F1. In the LCD only the Phrases that lie directly over the →Position Line can be used for this. Making the choice is carried out either with the VS's Track Status buttons or using the Cursor buttons and F3 MARK. This is indicated by a checkmark in the Checkbox in front of the Name. All Phrases above the Positions line can be chosen in one step with F2 ALL or can be deactivated together by pressing F2 again. Each Phrase in the VGA monitor, however, even if it lies outside the area of the positions line, can be selected with one Mouse click

Tip:
Chosen Phrases can be deactivated simply by pressing F2 (ALL) twice. All Phrases are otherwise activated by using F2.

Send
1. Send/Return effect
In →Send/Return Effects, both the procedure for transmitting signal parts to an →AUX Bus and the display knob are called S.

2. Insert effect

Not to be confused with the Send/Return effect described below, an →Insert effect in the VS has controls with the same name available to it (LCD labelling, not in the VGA). This then levels the volume before and after the Insert effect in a channel or in the Master Section.

Send/Return Effect
A S. (also known as a Loop effect) represents a typical connection for many classic →Effects such as Reverb, Delay, Flanger etc. and can be activated in the VS without any additional Routing.

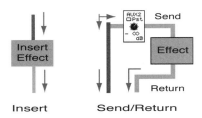

Fig.: Insert and Send/Return effect

The way a Send/Return effect works is that it mixes Effect Signals (wet) into the untouched Original Signal (dry), whereby the mixing relationship is determined by the respective Send knob. An unlimited number of Send signals can be fed from the Channels of the Input and Track mixers and from an →FX Return Channels via →AUX Bus* into the effect device for editing. An →Insert can, however, only be used for one Mixer Channel.

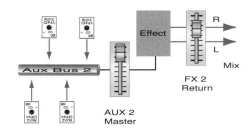

Fig.: Block diagram of a Send/Return effect.

The stereophonic Effect Return Signal ("Return") from the Effect Processor is leveled in the →FX Return Channel, edited and then sent in the →Mix Bus for listening. Depending on the number of installed →Effect Boards, multiple Effects can be used simultaneously (e.g. in the VS-2400: FX1 for Reverb, FX2 for Delay, FX3 for Chorus and FX4 for a guitar effect). Additionally, →External Effect Processors can also be integrated in the VS via Send/Return.

Calling up the CH EDIT page
Generally setting a Send/Return is limited to the leveling of the respective →AUX Send knobs in the Mixer Channel. (By default both of the global Parameters →Aux Master and

→FX Return Channel are set to 0 dB. Where applicable, these Settings should be checked.) To edit an Aux Send value, the Mixer Channel's →CH EDIT page is normally called up. Pay attention here to the appropriate Mixer Section (→Input Mixer and →Track Mixer respectively posses two sections in the VS-24xx).

VGA:	Double click on the Track Number
VS:	CH EDIT button in chosen Section
VS-2000:	CH EDIT → round Track button

VS-2000:
Firstly assign the 16 buttons to the "Channel Edit" mode using the CH EDIT button. The buttons' LEDs are thereby no longer lit.

Leveling Aux Sends
Each Mixer Channel provides an Aux switch, a Send knob and Pre/Post switches for all of the 8 Aux Busses*:

Fig.: aux send knob in the VGA

Be certain that the respective Send switch is activated (yellow). Then, using the Send knob located underneath, the signal can be sent to the Aux Bus and therewith into the associated Effect Processor. The effect then becomes audible. The reverb most used in practice is already assigned to the AUX1's effect processor in the default Settings.

  * *In the VS-2000, instead of "Aux" for the internal Effects, the indication "FX" is used.*

Aux Sends can also be leveled in the VGA menus →TR F/P, →MltChV with the Mouse.

Pre and Post
Generally, the Signal is sent after the Fader (post). This maintains the same Effect/Original ratio independently of the Channel volume, which is mostly aimed for. For special applications (e.g. Fading from 100% Effect up to 100% Original Signal) the Setting "Pre" can be used (→Pre/Post).

Note
The Pre/Post settings in one Mixer Channel affect all channels of the same Aux Bus.

Level Meter
Not only the →Aux Master levels but also the →Effect Return levels can be monitored using the →Level Meters AUX/DIR or FX1-8 RTN. The following table shows the Level meter indications for the corresponding sections:

Section	VGA	F button
Aux Master	AUX/DIR	F4
FX Return	FX1-8 RTN	F5
		(PAGE 2/3)

Fault clearance
In case the desired Effect is inaudible, the signal flow can be controlled with the help of Level Meter:

1. Signals sent to the Effect Processor with Level Meter AUX/DIR.
2. FX Return Signals sent out of the Effects with the Level Meters FX1-8 RTN.

Sensitivity Knob
An S. is part of the →Microphone Preamp (p. 122). It determines the degree of the signal intensity to optimally fit with the level for the →A/D Conversion. Generally →Level Setting While Recording (p. 104) is only carried out via S. (Set the Faders in the →Input Mixer (p. 97) to 0 dB).

Sequencer
An S. serves to record, playback and edit Audio and Midi data in a Computer that has the Sequencer software installed on it (Software sequencer) or in a Device meant for this purpose (Hardware sequencer). A VS is therefore also an Audio Sequencer. In the VS-2480, →Phrase Sequence (p. 155) allows editing of Sample Events. In the VS-2000 S.s are also available for the functions →Harmony and →Rhythm Track.

Shelving

S. Filters are parts of an →Equalizer and provide a simple High and Low boosting or lowering. Here the Parameter F (Frequency) changes the point at which the →Filter is operating. The parameter G (Gain) cuts or boosts the frequencies below F. (Low Shelf) or above F, (High Shelf). In the diagram below the low frequencies have been lowered and the high frequencies boosted:

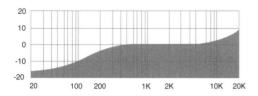

Fig.: Shelving EQ

Whilst the Shelving EQ is used for smooth lowering or boosting of frequency ranges (slope = 6dB/oct), the High or Low pass's 12~24 dB slope limits the frequencies almost completely in one direction (only in the VS-2480).

Shift

Because of the restricted number of buttons on the VS, certain operations take place by holding down the SHIFT button. This double function is indicated by a black frame on a white background whilst a label without this frame refers to the first function. In this way, the ZERO button positions at the beginning of the Project whilst SHIFT + ZERO (STORE) activates the saving procedure.

In the VGA with the help of SHIFT, clicking on the corresponding Icon generally calls up a Menu directly.

Example: Clicking on [♪] activates the metronome. SHIFT + Clicking on this Icon, however, calls up the metronome Menu.

Tip
The SHIFT button can be set for a single handed operation (e.g. Guitarists) using →Shift Lock.

Shift Lock

Particularly for Guitarists/Bassists who are themselves playing, calling up the second operation of a VS button via →Shift with both hands can be rather uncomfortable. Using the S., the SHIFT button can therefore be locked for one single operation or permanently:

Utility → Global1 → *Shift Lock*

Once
After pressing SHIFT, the "shift" function remains active until the next button is activated. This recommended mode of operation allows calling up the double function with only one hand:

1. Press SHIFT (LED blinks)
2. Activate the desired button for the double function. This function becomes active, the SHIFT LED ceases.

On
The SHIFT button is activated permanently by pressing it once and will be freed again only after firstly pressing SHIFT again.

Off
The normal operation requires simultaneously holding SHIFT to call up the double function.

Shutdown

The S. operation serves to correctly shutdown the system and also to position the Write/Read heads in a safe standby position.
Merely turning the VS off can lead to hard disk damage and data loss.

SHIFT + STOP

VS-2000
For this operation, the VS-2000 provides a dedicated SHUTDOWN button, which is to be held down for at least two seconds for operation.

VGA
All models can be correctly shut down via the project menu in the VGA.

Project → Shutdown/Eject

SI-80SP

The optional →Synchronizer S. serves to establish a frame synchronization between VS Recorders (0r →MIDI Sequencers) and digital Video Recorders/Camcorders using →MTC. Not only the sony LANC, but also the Panasonic Edit E protocols are supported.

Fig.: Video Synchronizer SI-80SP

The →Timecode of the Video device in addition to playback is also transmitted in slow motion, when increasing/decreasing Frames and whilst fast forwarding/rewinding (FF/REW). This is suitable for post production. The required setting is limited to the →Frame Rate selection. The Field Recorder →R-4 (Pro) from Roland's subsidiary company EDIROL provides an internal LANC/MTC converter.

Side Chain

In →Dynamics, the analyze signal is normally identical to the signal being processed. In case another signal should act as the analysis signal, it will be fed into the Dynamic's Side Chain Input (→Key In in the VS).

Signal Noise Ratio

The difference between wanted signals and given noise level, created by the device itself is represented by S. and is shown in →dB.

Sine Wave

The →Generator/Oscillator of the VS-2400/2480 provides a tunable sine tone generator for monitoring and checking purposes. Furthermore →White Noise, →Pink Noise and the →Metronome sound can be outputted.

Skip Project Load

In the worst case, the partition's TOC can be faulty. The VS hangs while loading the last used project and there is no other possibility to switch off - without carrying out the shut down procedure. As a last resort in order to rescue the remaining projects, S. enables a startup procedure with a project list exclusively for loading the last (damaged) project:

Hold Track Status 4 and CH Select 4 → Switch on

After a few seconds a message appears:

"Boot Condition: Skip Project Load"

In this state any partition can be selected via F4 (List) and the desired projects can be loaded then as back ups. The faulty project should not be selected. After this hazardous operation the harddrive must be →Formatted.

Slave

With the →Synchronization of two devices, the Slave analyzes the →Timecode sent by the →Master whereby the synchronization is operated with its time reference.

SMF

"Standard Midi Files" represent standardized data for playing music in Midi Sound modules and can be therefore played on any →MIDI Sequencer. Because they are all purpose and can be used universally, VS users can share settings using SMF (→MIDI Bulk Dump). In the Rhythm Track of the VS-2000, S. can be imported from a CD Rom (→3.6. SMF Import for Rhythm Patterns (p. 197)).

SMPTE

The SMPTE Timecode (Society of Motion Pictures and Television Engineers) was developed for an exact frame →Synchronization of picture and sound. It subdivides time values in Hours:Minutes:Seconds:Frames:Subframes (→Frame Rate) and is recorded as an analog signal on a track of a Multi Track Recorder or an analog Video Recorder (LTC). For Midi applications SMPTE is transferred into →MTC. Out of all the VS models, only the VS-2480 can be externally controlled directly with SMPT without an additional →Synchronizer:

Snapshot

VS-2480: Utility → Sync Parameters:
Ext Sync Source = SMPTE IN

The →Time Display of the VS is also described as an SMPTE indicator.

Snapshot
→Automix (p. 19)

Soft Knee
The S. characteristic of a →Compressor allows an inconspicuous compression even when ratio values are high by using a curve instead of the abrupt angle of a normal hard knee. This is used smoothly just before the Threshold Level and controls the relationship of the compression depending on the degree the threshold is exceeded. The Compressor of the Roland Plug-In →Vocal Channel Strip provides a switchable S.

Solo
The Solo function of a Mix table allows one or more of the chosen mixer channels to be listened to in isolation. It is achieved by muting the remaining channels and serves to control one channel signal without having to change the corresponding Fader Settings. By operating on the VS panel, the CH Edit buttons serve as solo switches. For that purpose they must be assigned via the superior SOLO button first.

VGA:	Click on a [S] in →Playlist, →TR F/P or →MltChV
CH EDIT	Click on [SOLO]
VS-2400, VS-2000	SOLO → Press a flashing CH Edit button
VS-2480:	SHIFT + SOLO → Press a flashing CH Edit button

Hint:
The VS's SOLO button doesn't serve as a main switch for the Solo function. The short cut CLEAR + SOLO, however, can deactivate the solo status of all Mixer Channels in one step. By activating Solo for a Mixer Channel, the message "SOLO" blinks both in the VGA's footer on the right (blue) and in the LCD's header.

Deactivating single Solos
On the VS panel, after pressing the SOLO button again (the LED is no longer lit), the channels activated in Solo status remain – This is shown by the blinking "SOLO" indicator in the LCD and the VGA. The VS's SOLO button only assigns the solo function to the CH Edit buttons and doesn't serves as a Solo main switch. Channels are, therefore, to be individually deactivated again. In the VGA this is carried out by clicking on the blue blinking Icons.

Deactivating Solos together
All Solo activated Mixer Channels can be deactivated together using a shortcut. The VS's SOLO LED, however, remains lit to indicate the button assignment to the solo function. Only after pressing SOLO again, which is shown by turning off the LED button, is the special assignment canceled and the normal mode re-established.

VGA:	Playlist: Click on the arrow in the Footer of the Solo Column
VS:	CLEAR + SOLO → Switch off SOLO (s.above)

Tip:
- The Solo status for all Input or Track Mixer channels is controlled simultaneously in the LCD via the menu →Parameter View (p. 142).
- The effect of the Solo function can be limited to the Mix Bus using the parameter →Solo/Mute Type.

VS-2000
Only the VS-2000 offers the handy option →FX Rtn SOLO ENABLE, which doesn't mute an →FX Return Channel when the Soloing of Mixer Channels is activated. By doing this, the channel signal can be listened to together with an assigned →Send/Return Effect.
LCD: The Mutes of all Mixer Channels can be

Speaker Modeling

deactivated or activated simultaneously. To do this, position the cursor on the Mute Field in any CH EDIT page. By holding CLEAR + turning the Time/Value dial, the desired Status for all Mixer Channels is chosen.

Solo/Mute Type
The functions →Solo and →Mute can be either limited to the →Mix Bus or expanded together to all →AUX Busses and →Direct Paths:

   Utility → Play/Rec Parameter → *Solo/Mute Type*

Particularly by using →AUX Busses for monitoring, Solo and Mute can be active either for the sound engineer or also for a headphone mix for the musicians.

Soundblender
This optional →Plug-in provides pitch shifter, chorus, Flanger, Phaser, dual resonant filters, echoes, revers echo and arpeggiator for thick and deep guitar and keyboard sounds.

Fig.: Plug-In"SoundBlender" from SoundToys

SoundBlender	
Effect Patch:	Plug-In
Connection:	Send/Return
Application:	Guitar

The intelligent pitch shifting allows scaled harmonies (common and ethnic scales) to be created. The original developers came from the Eventide company, which is known for the first studio harmonizers (H 3000).

Source
In general, S. describes the origin or the source of an operation that is to be carried out (→Bouncing, →Track Import, →Region Edit, →Phrase Edit). →Destination (p. 54), on the other hand, represents the destination point or target.

Space Chorus
→Chorus (p. 45)

Speaker Modeling
S. simulates the sound behavior of popular studio monitors with the use of default reference loud speakers. By comparing a reference recording with one's own mix respectively using different sounding monitors, an unbalanced frequency spectrum can be found more easily. Even though at the time of printing only the Roland DS-90/50 (no longer available) are designed to be reference monitors, the →Effect can also be sensibly employed with other monitors in →Mix and →Mastering for an →A/B Comparison. To achieve completely different frequency responses, it is crucial to control the spectral balance and that not only one certain speaker is used.

Speaker Modeling	
Effect Patch:	P 034
Connection:	Stereo Insert
Application:	Mix, Mastering

Output Speakers:
Using the (no longer available) Roland DS-90 and DS-50 Monitors, represents the exact simulation of the Speakers to be found under "Model". By using other reference boxes, the result will not be correspondingly achieved but nonetheless allows a sensible use to compare Mixes (See above).

Model:
In addition to the simulations from Genelec 1031, Yamaha NS-10 Auratone etc. a linear frequency response can also be achieved with "Super Flat".

Split

Tip
- It is recommended to intermittently cross check the simulations of the mini speakers (Small Cube/Radio/TV), which furthermore realize a mono connection (→Mono Compatibility).
- The sound changes, used in the post-production of a playing radio when leaving the room, can be easily achieved by changing to "Small Cube" (switch →Bypass on/off).

In order to adapt the speaker sound to the specific room conditions the following blocks can be used:
BCut: →High-Pass (cuts low frequencies)
LFT and HFT: Low/High Frequency Trimmer (→Shelving)
Lmt: →Limiter

Split
→Phrase Split (p. 156)

Split Point
In the →Mastering Tool Kit's 3 Band Expander/Compressor, the frequency band ranges for separate editing can be specified using the two Split Points. In this way for example, the Split Points of 300 Hz and 3 Hz then generate the following ranges:

Low-Band:	20 Hz -	300Hz
Mid-Band:	300 Hz -	3 KHz
High-Band:	3 KHz -	20 KHz

SRV Stereo Reverb
The →Plug-in S. is delivered along with the Effect board →VS8F-3. It offers a higher Reverb quality than the Algorithms of the internal →Effect Board (up to 96k) and as a →True Stereo Reverb is able to create room for Stereo Tracks whilst still maintaining the R/L positions.

SRV Stereo Reverb	
Effect Patch:	Plug-In
Connection:	Send/Return
Application:	optional

S. additionally provides a →Compressor and an →Expander for processing the signal that is sent to the Reverb. The graphic display of the Early Reflections and the diffuse reverberation in the VGA offers a good overview and makes assigning the controls to the corresponding parameters easier.

Reverb
These Setting parameters correspond to those described in →Reverberation (p. 193). Additionally, the L/R balance of the Effect Signal can be specified.

Compressor
The vintage →Compressor reduces the Dynamic of the Signal that is to be reverberated and creates through this a compact and thick Reverb effect. Using Type, either Transistor (Solid) or different tube characteristics can be simulated. A →Soft Knee curve (always activated for Tube) can be connected for a smoother Compression.

Expander
The →Expander controls the Effect Level depending on the volume and length of the Signal sent to the Reverb. Whilst "Threshold" attenuates Signals below this level, using the Parameters Attack and Release specifies the length of time the Effects are exposed.

St Flanger
This type of →Flanger provides →Effect editing of Stereo signals. An →EQ is located after the Flanger for further sound editing.

Stereo Flanger	
Effect Patch:	P 100, 101
Connection:	Send/Return
Application:	Standard

St Phaser
This →Phaser allows effect editing of a Stereo signal* (→True Stereo) with up to 16 stages. It can be used for Guitar chords, pianos and also for Solo Instruments by creating a pulsing sound.

Stereo Phaser

Effect Patch:	P 097, 098
Connection:	Send/Return
Application:	Standard

* "Pol": Set Mono signals to "Inv"

Stage Depth
→Mix (p. 125)

Start
To navigate to the first recorded portion on visible V-Tracks see "Top" p. 225.

Status
→Track Status (p. 229)

Stay Here
In the →Mastering Room's "After Rec", the Setting "Stay Here" produces an alignment of Master and the Recording Tracks. Through this, the level of single instruments (!) can be easily changed subsequently to →Mixing (→4. Correcting A Mix Without Redoing It (p. 127)).

StDly-Chorus
This combination of →Delay, →Chorus and 3-Band →Parametric Equalizer can be used to produce a spatial →Effect (broaden the sound).

Stereo Delay Chorus

Effect Patch:	P082-084, 91f
Connection:	Send/Return
Application:	Standard

Stereo
Stereo Signals are recorded and edited preferably in linked Mixer Channels in the VS. For this, either the →Channel Link, which changes all Mixer Parameters of both channels simultaneously, or the →Fader Link which only links Faders, are available. Both modes can be activated in the respective →CH EDIT menus. Particularly for →Recording, the corresponding Channels in the Input and Track Mixers (→Mixer) are to be coupled using →Channel Link for a stereo recording in order to ensure the R/L assignment. Both the →S/PDIF sockets of all VSs and the VS-2000's Track Channels 15/16 and 17/18 are provided in stereo.

Automix
When editing →Automix data, the volume or the →Panorama data of linked channels is saved as →Offset Level or →Offset Pan.

Stereo Mix
The result of mixing is generally a Stereo →Mix. In addition to playback tracks, also →Effect Returns signals and, when applicable, sound synchronized Midi modules from the Input Mixer go into this. The S. consists of two channels (R and L) and is recorded as →Master Tracks. These create the basis for the CD that is to be created. In the →Surround mix however, up to five separate channels can be generated.

Stereo Multi
This Patch (→Effect) provides in stereo:
A →Noise Suppressor (to suppress background noise), a →Compressor (attenuates signal levels when reaching the Threshold Level), an →Enhancer (Level dependent boosting of high frequencies) and a →Parametric EQ (to boost or cut individual frequency ranges). Because of the high quality of the Mixer Channel's →Dynamics and →Equalizer, generally only the Enhancer and the Noise suppressor of this Patch are necessary. The →Mastering Tool Kit offers a better alternative.

Stereo Multi

Effect Patch:	P 033
Connection:	Insert
Application:	Special

Store
In regular intervals, after recording and other important operations, always save the current project (→Project Store).

SHIFT + ZERO (STORE)

Store Cur rent?

This prompt appears before a Project, CD or other operations can be carried out and refers to the necessity of saving Settings that were made. Normally the answer here is YES. If audio recordings, Edit Operations, changes to the Mixer and system settings etc. should *not* be adopted, continue with NO. This security prompt only doesn't appear when protected Projects are being used (→Project Protect).

StPS-Delay

In addition to conventional delays, experimental →Effects can also be created with the S. It provides a →Pitch Shifter, a →Delay and a 3 band →Parametric Equalizer for sound design.

Stereo Pitch Shifter Delay	
Effect Patch:	P004 +104
Connection:	Send/Return
Application:	Special effect

Studio Monitors

Contrary to those of music, mixing and mastering set other demands on Monitors and Listening environments. Ideally, the Studio monitors should provide a linear frequency response and a high impulse consistency in order to play back the sound spectrum being judged neutrally - Hi-fi boxes generally produce a cut in the mid range to sweeten the sound. Due to technical reasons however, a linear frequency response is not to be achieved so that a Monitor with boosted mids is given precedence instead.

Tip:

- The sound of Studio monitors can be changed using →Speaker Modeling.
- When working in a small studio room, the low ends of a mix can be better judged in an adjacent room.
- Using an Equalizer in the Monitor signal, the Frequency response can be corrected.

SUB DISP

In →Linked channels (p. 104), single levels and pans can nonetheless be individually adjusted (→Offset Level (p. 137), →Offset Pan (p. 138)).

Subframe

An S. represents the 1/100 fraction of a Frame (→Frame Rate) and therewith corresponds to 0,33 ms with 30 fps and 0,42 ms with 24 fps. In the →Time Display of the VS, the Subframes field is located between Frames (f) and Measures (Meas).

Subgroup

Joining multiple Mixer Channels to an additional channel for combined editing is called a S. Generally group channels are thus collectively leveled, filtered, compressed or edited with a bus effect. Even though S.s are not provided in the VS, they can be created through special routings:

Fader groups

Moving just one fader of a →Group allows the level of several mixer channels to be set. Further editing of the Group Signal (e.g. EQ, Compressor) cannot, however, be carried out.

Routing to a Recording Track

Routing the desired channels to a Track (often a stereo track) that is on →Record Standby, achieves outputting exclusively over this track. The steps of this procedure are the same as with Track →Bouncing. The condition for this is that the →Record Monitor parameter is properly set:

Utility → Play/Rec → *Record Monitor = Source*

For the sake of simplicity, "Source" is to be selected here.

Aux Bus

Using stereo coupled →AUX Busses, signals can be positioned freely in the stereo field. After →Linking two Aux Busses, deactivate the →MIX Switches in the desired Mixer Channels The signal is then fed into this Bus via →AUX Send and the Panorama setting is carried out using the Pan knob of the linked Bus. Following this, the Aux Bus is fed again into the →Input Mixer (!). Here, the desired editing in the form of a real subgroup is carried out with →CH EDIT Parameters. The following methods are available for Routing into the Input Mixer:

Analog
In →EZ Routing (Output Assign), the Aux Bus is routed to analog outputs (→Output Assign). Connect the AUX OUT jacks with two Input sockets via guitar cables. Then link the Input Channels.

VS-2480
The digital →S/PDIF output in the VS-2480 can be connected to the S/PDIF Input. No →Clock settings need to be carried out for this. Here the Aux Bus is to be routed to COAX out (or Optical out) in the EZ-Routing's "Output Assign". In the Patchbay, the connection to COAX In (or Optical in) of the desired input pair is to take place.

R-Bus Routing (VS-2480)
Connecting both R-Bus sockets in the VS-2480 allows additional Routings, which cannot be achieved conventionally (→R-Bus Patching). In this way, real Subgroups are created, which are edited in the Input Mixer's Channels.

Subsonic Filter
An S. serves to suppress low frequency noise such as subsonic noise (via mic stands), impact sound (via microphones), humming, buzzing etc. In the VS, a →High-Passfilter at approx. 80 Hz (or below) can be used in each Mixer Channel (except Bassdrum and Bass)- especially for →Live Mix and →Live Recording.
This is only available in the VS-2480 as a separate Filter, so that it helps by sinking the lower frequencies using a Low →Shelving Filter.

Subwoofer
A Subwoofer is an additional speaker cabinet, which only transfers the low ends (below 80 Hz) and therefore rounds the sound characteristic of smaller boxes up. Often in →Surround, the LFE channel (Low Frequency Extension) is incorrectly described as a Subwoofer.

Surface Scan
During the →Formatting of the hard drive, a surface check for mechanical damage can be carried out in the form of a Write/Read Test. Damaged sections are thus excluded from any further processing.

Surround
The VS-24xx models are able to carry out surround mixing for the formats 2+2, 3+1 and 5.1. Through this, Signals within the field created by at least four loud speakers can be freely positioned. Whilst Stereo Panning only allows positioning between L and R, Surround Panning additionally enables movements to the Rear speakers Ls and Rs (Left/Right Surround). For the 5.1 format, the recommended speaker set is shown in the following diagram.

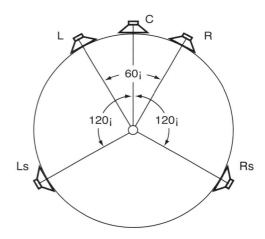

Fig.: Positioning the Monitors for 5.1 Surround

In 5.1 format, an additional LFE channel* (Low Frequency Extension) is provided, which only transfers low frequency effects in the Sub base area.

  * *Low ends can be separated by the →Isolator, 3-Band or via a VS-2480's →Low-Pass.*

Static positioning and levels are programmable as normal in →Scenes and dynamic changes using Automix.

Surround

Hint:
The VS can be used for internal recording of surround channels in addition to surround mixes. This is not, however, intended for encoding in the AC3 or DTS format for playback of conventional DVD players. This remains a designated function of the respective Hardware (e.g. Dolby®), Software Plug Ins or special Surround Studios. For surround playback, the VS uses regular →AUX Busses, which are then no longer available for further applications. Therefore when, for example, in the Surround Format 5.1, only the Aux Busses 1 and 2 remain.

Utility → Surround → Surround Mix Sw = On

A prompt refers to re-configuring the Aux Bus and the Panorama assignment. Depending on the actual existing number of loudspeakers, one of three formats is then to be chosen. The Surround Channels are therewith concerned with the following analog outputs:

Format	L/R	Ls/Rs	C	LFE
2+2	5/6	7/8	-	-
3+1	5/6	8	7	-
5.1	3/4	5/6	7	8

In the default Settings, the respective analog (and R-Bus) outputs correspond to the Busses with the same numbers. The assignment of these can be changed in the →EZ Routing OUT or in the →Master Section.

Surround-Panning
When Surround mode is activated, the VS adds the Surround field to the →CH EDIT page. In this field, the Surround Panning can be carried out using the Mouse*. Movement of the red ball in the Display positions the channel signal in the real loudspeaker field. This is realized by sending the corresponding Signal levels to the respective Aux Busses, which are easily visible in the LCD's Level Meter AUX. In this way, moving the ball down (Ls and Rs) automatically produces the signal output in those Aux Busses,

which are connected with the back surround loudspeakers. Center and LFE channels of the 5.1 Format are independently leveled with the LR:C and Sub.W (Subwoofer) controls.

* *A useful tool for Surround panning is the Joystick of the optional Channel Strip →VE-7000.*

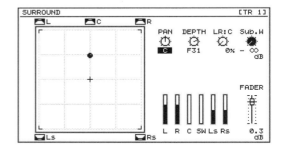

Fig.: Surround field in the LCD

Automation
Static positions of channel signals can be stored via Scenes or Automix Snapshots.
Surround movements are fully integrated in the Automix. To do this, surround has to be activated for →Automix:

Utility → Automix → Setup → Writing Parameter

The →MltChV page concisely displays Surround Positions and movements of multiple Channels and furthermore allows an active S.Panning.

Effects
For using Effects in surround, activate the aux send switch (L, R, Ls, Rs etc.) in the respective →FX Return Channels. These can then be also positioned in the room. It is recommended to provide Reverbs for the front and rear. However, this doesn't substitute a Surround Reverb (e.g. Roland's RSS-303).

Leveling in Surround Mode
When operating in Surround, the Master Fader is not available. The volumes of all the Bus signals for surround playback are controlled by the respective →Aux Master Faders.

Sync Track

Recording the Surround Bus Signals
The bus signals of the Surround Channels (6 channels for the 5.1 Format) can be recorded to empty tracks. Because the Bus Outputs appear in the →EZ Routing's "Track Assign", they can be routed to the desired recording tracks (→Bouncing) and recording can then take place.

Tip
For encoding in a PC or in a Surround Mastering Studio, the recorded Surround Tracks can be exported via →Wave Export.

Switch off
To correctly close a project and to subsequently turn off, the →Shutdown procedure of the VS Recorder should take place. By doing this, the last recordings and Settings will be saved, the Write/Read head of the harddrive will be set at a safe standby position and the harddrive will be conventionally shut down. Simply turning off using the power switch can lead to data loss and/or damage to the Harddisk.

VS-24xx: SHIFT + STOP (Shutdown)
VS-2000:   Hold SHUTDOWN button 2s

Switching Time
The reaction time for calling up the →Quick Routing menus by holding one of the STATUS buttons is determined by the Parameter Switching Time with a range from 0,3-2 seconds:

Utility → Global1 → *Switching Time*

Sync Auto
During →Synchronization as the →MTC Slave, the VS with an activated "Sync Auto" automatically recognizes the received →Frame Rate. In this way, manual setting is then no longer applicable (not in the VS-2480).

Utility → Sync Parameters → *Sync Auto* → *ON*

Sync Offset Time
In the →Synchronization, when a VS acts as the Master, only the transmitted →MIDI Clock can be delayed. By doing this, an →Offset can be achieved.

UTILITY → SYNC Prm → Sync Offset Time

Sync Parameter
In the "Sync Parameter" page of the Utility menu*, the Settings necessary for →Synchronization will be carried out.

Utility → Sync Parameter
VGA: Click on

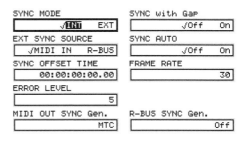

Fig.: Sync Parameter

In the VS-2480, the Sync page can be directly called up via SHIFT+EXT SYNC.

Sync Track
In the S. the Midi Clock Data of an external device can be recorded. In contrast to →MTC, the tempo is contained within the →MIDI Clock (p. 123) data. Using the S., Tempo changes can be applied with a higher resolution than that of the internal →Tempo Map (p. 223)'s measures. Tempo changes of an imported Midi Clock can be saved and then used to synchronize an external →MIDI Sequencer and for the VS' s →Measure Display. Via S., imported Tempo changes take place gradually in the VS, if it is acting as a Master in the Midi Clock →Synchronization (p. 217) and for it's own Measure Display. For each project, one Sync Track can be saved.

Hint:
A Sync Track can also be created manually (→Sync Track Convert (p. 216)).

Sync Track Convert

	Sync Track	Tempo Map
Resolution	1/96	1 Measure
Creation	Ext. Sequencer	VS
Editing	no	yes

Hint:
Tempo and Tempo changes don't have an affect on the recorded Audio data. They do however influence the Measure indicator, the metronome, the internal →Rhythm computer of the VS-2000 and the transmitted tempo via Midi Clock. The easy →Editing in a rhythmical →Grid remains possible even with tempo changes in the Project.

1. Recording the Tempo Information
After establishing the →MIDI Connections, activate recording in the VS for the Sync Track:

UTILITY → SYNC Prm → F1 STrRec

By starting the Midi sequencer, the VS records the Tempo Information contained in the Midi Clock whereby *"Now Sync Track Recording"* is displayed: After receiving, the recording of the Sync Track is closed with *"Done"*.

2. Playback using the VS as the Master
In the Sync Menu, the entry "SynTr" is to be selected for the Sync Generation Parameter.

UTILITY → SYNC Prm:
MIDI OUT SYNC Gen. = SyncTr

After starting Playback in the VS, the measure indication changes correspondingly and the Tempo changes are sent via Midi Clock Data to the external sequencer.

Activating/Deactivating the Sync Track
By choosing another entry other than "Sync-Tr", the Sync Track is deactivated and the Settings of the Tempo Map become active again.

Sync Track Convert
The Tempo, including tempo changes of a freely (without a metronome) recorded song can also be subsequently determined. To do this, the function S. converts Marker Positions or a defined time section either to a new →Sync Track or a →Tempo Map. (A Tempo Map can also be created from an imported Sync Track.) Whilst the →Sync Track is generally only needed to synchronize an external →MIDI Sequencer via →MIDI Clock, the Tempo Map represents the basis for a correct measure indication in the VS itself.

Utility → Tempo Map → Tm Cnv (F1) →
Convert Type:

The following conversion types are available:

1. Deriving the Tempo from Markers
"Tap → Sync Trk" or "Tap → Tempo Map" respectively generate a Sync Track or a Tempo Map. Firstly, according to the recorded music's tempo, →Markers are tapped in. Here beats or half notes are recommended. When the song contains Tempo changes markers have to be tapped in till the end of the Song. This can be carried out either in realtime and/or (very accurate) with the help of the →Waveform Display. Correcting the marker positions is possible at any time. The BEAT used and the number of markers tapped in per measure (TAP BEAT) must be entered into the corresponding fields. Using ENTER carries out the conversion so that either the new Sync Track or the new Tempo Map are then available.

A Tempo Map (for VS internal use) can be directly converted from a created Sync Track via "Sync Trk → Tempo Map". In this way for example, imported Tempo changes from a Midi sequencer (→Sync Track) can at least be crudely adopted in a Tempo Map (the Tempo is only able to be edited measure for measure).

2. Tempo From a Predefined Time Segment
The setting "Time → Sync Trk" allows the Tempo of a predefined Time Segment to be specified. To achieve this, the number of measures and the beat type (e.g. 3/4) are to be entered in addition to the Start and End time of the Passage. The VS then calculates a Sync

Synchronization

Track with the help of this parameter. Using "Sync Trk → Tempo Map", a Tempo Map for the proper VS measure indication can be generated again.

Tip:
The function *Time → Sync Trk* can be used as a practical calculator for determining the Song tempo from the elapsed time of a passage.

Utility → Tempo Map → F1 (TM Cnv)

Sync with Gap

In →Synchronization mode as Slave, an activated S. sustains the synchronization even when there are small drop outs in the →MTC. This is also called Jam Sync.

Utility → Sync Parameter → Sync with Gap

Combined with the →Error Level Parameter, either a higher security or a higher accuracy can be achieved for the synchronization.

Synchronization

S. determines the process of locking two recorders together, so that they operate as an device. The so called Master sends the synchronization data and the Slave, using this data, aligns itself exactly to the Tempo provided by the Master. In practice, the VS can be used equally as Master and Slave. This results in the following applications:

VS - Midi sequencer
VS - VS
VS - Video recorder*
VS - DV-Camcorder*
VS - Analog Tape Recorder*

An additional →Synchronizer is required

In the VS, S. is realized using Midi or →R-Bus connections (VS-2000 uses only Midi). After carrying out the particular →MIDI Connections, depending on the desired operation mode, different settings in the Midi and also the Synchronization Menus (→UTILTY Menu) are to be carried out.

1. VS as Master

This mode of operation is generally used in connection with a →MIDI Sequencer. Either →MTC or →MIDI Clock can be used as Synchronization signals depending on the ability of the Slave. Whilst Midi Clock represents an adequately accurate, uncomplicated to deal with mode and above all contains the Song Tempo, the more accurate MTC demands numerous Settings in Master and Slave.

1.1. Midi settings

With an MTC Synchronization, pay attention to the control of the transport functions, which are realized via →MMC. No Settings need to be carried out for the Midi Clock (here MIDI OUT is a default setting).

UTILITY → MIDI PARAMETER:

Midi-Parameter	MTC	Midi-Clock
MIDI OUT/THRU	OUT	OUT
SysEx.Tx Sw	On	-
MMC Mode	MASTER	-

1.2. Sync settings

For the MTC operation, the correspondence of the →Frame Rate in the Master and the Slave must be paid attention to in order to avoid the device's times drifting away from each other. When using a Midi Clock Synchronization, only the MIDI clock needs to be activated. In special cases for a Midi Clock Synchronization, a →Sync Track* is available.

Sync Parameter	Master	Midi-Clock
SYNC MODE	INT	INT
Midi Out Sync Gen	MTC	MIDIclk
Frame Rate	as per Slave	-

The Sync Track plays back the recorded Tempo changes of a Midi sequencer. The limitations of the →Tempo Map can be then be avoided:

UTILITY → SYNC Prm:
VGA: Click on

Synchronizer

VS-2480:
The Sync menu can be also called up here directly via SHIFT + EXT SYNC

1.3 Offset
Only the outputting Midi Clock can be delayed using the →Sync Offset Time Parameter. Through this, an offset between Master and Slave can either be leveled out or created.

2. VS as Slave
The VS can be controlled by an external Master via MTC/MMC. Midi Clock cannot be used here.

2.1. Midi Settings
Transport functions for an MTC Synchronization are realized using →MMC.

Utility → MIDI Parameter

Midi-Parameter	MTC
SysEx.Rx Sw	On
MMC Mode	SLAVE

2.2. Sync settings

Utility → SYNC Prm:

Sync-Parameter	Master
SYNC MODE	EXT
Frame Rate	as per Slave

2.3 Offset
The time offset for the Slave playback delays the received Timecode. Un-synchronized Audio Passages from Master and Slave can subsequently be brought into correspondence using this (→Offset). The received MTC can be shown in the Slave either absolutely or under consideration of the offset (relative).

Utility → Project Parameter*:
"Display Offset Time "/ "Time Display Format"

* *VS-2000: "Display Parameter" menu*

3. Synchronization via R-Bus
Because the →R-Bus transfers not only Audio, but also Synchronization and Clock Data, one single R-Bus cable can be advantageously used for this. The Midi Settings correspond to the synchronization via MTC and MMC described above.

UTILITY → SYNC Prm:

Sync-Parameter	Master	Slave
SYNC MODE	INT	EXT
EXT SYNC Source	-	R-BUS1
R-BUS1 Sync Gen.	MTC	-
Frame Rate	as per Slave	-

Synchronizer
A suitable S. for the VS converts →Timecode into →MTC. This allows the →Synchronization of Recorders without a Midi connection with the VS*. Generally this concerns the →SMPTE and →Video Recorder Synchronization. The VS-2480 can be externally controlled using the integrated SMPT Entry

SysEx
→System Exclusive Data (p. 218) represents →MIDI messages, which are used for the manufacturer's own Editing and Data Dumping.

SysEx.Rx Sw / SysEx.Tx Sw
The Midi Communication using →System Exclusive Data can be separately turned on/off for Sending (Tx) and Receiving (Rx)

Utility → MIDI Parameter → SysEx. …Sw

The following operations require activation: →Synchronization, →MIDI Mixer Settings, →MIDI Bulk Dump etc.

System Exclusive Data
In contrast to the →Control Changes, the S. represent manufacturer specific non standardized Midi Data. Whilst only the Start and the End Bytes are defined, the remaining content can be used freely from the different companies. They

mainly serve to archive data of Sounds and Settings and to transfer realtime data. The VS allows all settings to be controlled. Make sure that the corresponding →SysEx.Rx Sw / SysEx.Tx Sw are activated for transferring data.

System Parameter

In the S. menus, functions and Settings are carried out, which concern the entire VS System and are used spanning whole projects:

Utility → System Parameter

The Parameters are generally split over Parameter pages 1-3, which are chosen via the corresponding →Function Buttons:
- →Phantom Power (p. 145)
- →Ext Level Meter (MB-24) (p. 72)
- →VGA Monitor (p. 236)
- →Mouse (p. 132)
  →Keyboard PS/2 (p. 101)

System Software

The S. represents the operation system of a VS and can be actualized and updated via →System Update. During →Booting the current Software version of the installed operation system is shown. Exact information about the System and Boot version is to be taken from the LCD's Version Page. To do this, before switching on, hold the following buttons until the Menu "Version Information" appears:

VS-24xx: STATUS 1 + SELECT 1
VS-2000:   STATUS 1 + HARMONY

System Update

Loading new →System Software in the internal →Flash ROM memory location of a VS Recorder is called a S. Through this, not only additional functions and Parameters become available but also Bugs (Program malfunctions) are repaired. This procedure affects neither the project data on the harddrive nor the project User settings in the VS memory locations (Routings, effects). The current System Software can be down loaded from the Roland US Website:

www.rolandus.com/support/downloads_updates
→ T-Z → V → VS-2xxx

The following methods are available for a system-Update: "CD-R", "SMF" or "USB" (Only in the VS-2000).

Via System-CD

A System CD can be written easily with a PC/Mac. Firstly, the corresponding file is to be loaded and unzipped. With the burning software Clone CD (PC) or Toast (Mac), an Image CD will be subsequently burned. Details can be found in the accompanying PDF Manual.

1. The following appears after turning on with the System Disk in the CD Drive: "Update Main System Program? Ver.2.xxx (Boot 2.xxx)" in the LCD.
2. YES calls up a security prompt referring to EZ-Routing and FX-User settings: "Keep User settings?"
   YES maintains these whilst NO achieves Initializing.
3. The Main Update is carried out. This procedure requires approximately 1 min. Then «Please Reboot OK» appears and the disk is ejected.
4. By turning off the power switch and turning it on again simultaneously holding Status 1 + CH Edit 1, the new system version can be checked.
5. ENTER boots the System.

Via SMF

A Hardware or Software sequencer is required for this. After unzipping, the SMF are loaded in the sequencer in the required order; The Sequencer's Midi Out is connected with the VS's Midi In via a midi cable. The VS then needs to be rebooted whereby the Status 2+3 and the CH Edit 5 buttons need to be held down. "SYSTEM UPDATE?" → YES → SMF consecutively transferred → "Update SysPRG?" → YES → "Please Reboot OK" → turn off.

T-Racks Mastering

Check up: When turning on, hold the STATUS 1 and CH Edit 1 buttons. The "Version Information" page then allows the new system software to be checked. YES starts the regular Boot procedure.

Via USB (VS-2000)
The operation Software can be transferred only in the VS-2000 via USB from a PC/Mac.

1. Turn off the Device, Don't connect the USB Cable.
2. Hold the USB button and turn on.
3. Connect the USB Cable. The VS Partitions appear on the PC Desktop
4. Copy the System Program in the first partition.
5. Log out of the VS Partition and extract the USB cable.
6. Press YES and the Update is carried out. Afterwards, the "Version Information" page appears.
7. Turn off the VS and turn it on again
8. Hold TRACK STATUS 1 + HARMONY and turn on in order to check the new Software Version.

T

T-Racks Mastering

Both the "Dynamics" and Equalizer →Plug-ins of the T Racks were created by "IK Multimedia" for →Mastering functions and are available as independent programs for Mac and PC.

T-Racks Dynamics

Effect Patch:	Plug-In
Connection:	Insert
Use:	Mastering

Because of the moderate computing required, only one Plug In can be operated in a Processor. For the Mastering procedure, EQ (e.g. FX1) is inserted before the Dynamics (e.g. FX2). Using the mouse is also necessary in the LCD.

1. Dynamics

Dynamics provides a compressor, followed by a Limiter. The →Bypass is active in the default settings and the Effect is engaged by clicking on F5 (Enable).

1.1. Compressor

The →Compressor (p. 47) uses it's "Drive" parameter to control the Input signal's volume. This corresponds at the same time to the well known lowering of the →Threshold level. With an extreme →Soft Knee characteristic, the compression is engaged early and softly, which comes close to the sound of classical tube compressors. The integrated Stereo Image Parameter is to be accentuated. This function, which is generally essential for Mastering, spreads the stereo basis above the normal measure and creates then the transparence in the sound image known from commercial productions.

Fig.: DynamicsModul of T-Racks

1.2. Limiter

The 3 Band Limiter effectively intercepts Peaks that are not suppressed by the Compressor. Using the Drive Parameter specifies the Limiter Input level even here (this is the same for the Compressor Out) and together with the Threshold knobs of the three bands, affects the degree of limitation. The maximum value of the Threshold knob is internally + 12dB. The limitation mode specifies the "Overload" Parameter. Whilst small values accord to the classic Limiting, higher values allow a larger level although this in turn increases Soft Clipping.

1.3. Output module
This module makes an additional Clipper algorithm available, which limits the output signal to a maximal level of 0.5 dB. The style of the Clipping can be specified from hard limiting where peaks are cut (knob fully right) to the smoother, more gradual Soft Clipping, which rounds up the slopes in the style of a tape saturation (knob fully left). This setting is to be specified experimentally depending of the material.

2. Mastering Equalizer
The EQ provides →Shelving Filters for the highs/lows and two tunable →Peaking Filters with two default →Bandwidths ( ). The amplitude of the chosen filter curve can change between 7,5 and 15 dB.

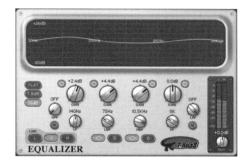

Fig.: Mastering Equalizer of T-Racks

In addition to the Shelving and Peaking filters, very effective →High-Pass and →Low-Pass filters (4. order = 24 dB) are also available.

T/DIF
The "Tascam Digital Interface" represents a digital interface with 8 channels. A 25 pin Sub D plug connector serves as the connection. Clock and MTC are fed over separate lines. Using →DIF/AT, the VS can be connected with a Tascam multitrack recorder.

Take
In the VS, a Take refers to an unchanged original recording, which is specified from the length of recording alone. Each new recording produces a new Take independently of its length of recording and recording position. In accordance with →Non-Destructive Editing the Phrases only represent copies in the form of pointer based references to Takes and can be edited at any time without having to change the original (→Undo). In the →Phrase New page, Takes can be managed, pre-listened and "reanimated" to Phrases and given a Name in →Take Manager or also irreversibly erased.

Take Manager
The T. essentially refers to the Take List from →Phrase New with sorting functions within a table for Name, Creation date or size and also a pre-listening option. It is not possible here to create phrases from Takes. Additionally, →Name Entry and deleting of Takes from the Hard disk can be carried out here.

Track → Take Manager

Using a Take in a Phrase is identified by a black box next to the Name. Unidentified Takes will be deleted from the Hard disk by carrying out the →Optimize operation.

Talkback
A Talkback microphone enables the Sound engineer to communicate with the musicians (→Headphone Mix). This microphone is to be exclusively routed to the corresponding →AUX Bus and be removed from the →Mix Bus. It's recommended to feed the T. Mic into the Aux Bus(ses) "Post Fader". Doing so allows a comfortable volume controlling with the respective Fader.

Tip:
- If all input jacks are completely occupied, an external →A/D Converter in the form of a Dat- Recorder (or something similar) and an additional microphone preamp can be used to help. Otherwise it is recommended to use both the last analog Input (8 or 16) and the Guitar Hi-Z switch for activation/ deactivation.

Tap

- The Talkback volume can be leveled in the Track Mixer via an unused Fader. This makes changing to the Input Mixer section unnecessary: to do this, a →Fader Group from the Talkback Fader in the Input Mixer and the Fader from the Track Mixer is to be created.
- Activating/deactivating the T. mic can be carried out by using the last microphone input via the →Guitar Hi-Z button. In the VS-2000 additionally via the →Tuner assigned Input.

Tap

Using the VS's TAP button or the VGA TAP button generates →Marker. SHIFT + TAP, however, calls up the →Tempo Map.

Tape Echo

The →Effect "Tape Echo" simulates a Vintage Tape Delay (RE-201) whereby the Echo repetitions are created using a combination of the three playback heads. Typical vintage Tape wow and flutters are integrated as setting Parameters.

Tape Echo	
Effect Patch:	P087 - P090
Connection:	Send/Return
Application:	Vintage

Target

For the display and correction of →Automix data the Mixer Parameter to be edited is firstly chosen via T. (→Selecting an Automix Parameter (p. 22)

TCR 3000 Reverb

The →Plug-in T. makes a premium →Reverberation Processor available for the VS.

TCR 3000 Reverb	
Effect Patch:	Plug-In
Connection:	Send/Return
Application:	Premium

It is to be connected in the form of a →Send/Return Effect. From the 100 Patches, one of the algorithms approximating the application can be chosen. This is then to be edited.

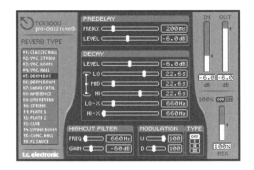

Fig.: Plug In TCR3000

Editing

The *Reverb Type* specifies the characteristic sound specifics in the form of a Macro. Ambience (spatiality with a small Reverb), Spring (mechanical), Room, Plate (mechanical), Reverb etc. are available here. In the DECAY field, the parameters for the diffused Reverb are specified. *Decay* determines the reverb time and *Level* the corresponding volume. The high and low frequencies can also be easily boosted or lowered here using *HI-C* and *LO-C* (contour). In the PREDELAY field, Reverb and Early Reflections are delayed with *Predelay* and the Volume of the ER is set via *Level*.

Finally in the Modulation Section via *Width* (W) and *Depth* (D), the speed and intensity of a frequency modulation for a diffused Reverb can be specified with different Types.

Template

As a master for similar recording and Mixing applications, for reasons of efficiency, project templates can be created without audio data but with ready made settings. When creating a new project, it is possible to access the Structure, Tracks, Mixer and other settings of similar projects. This saves having to recreate repeated settings:

Tempo Map

1. Creating a new Project (→Project New)
2. Carrying out the generally applicable Settings such as →Name Entry for the Tracks, →Routing, →EQ, →Dynamics etc.
3. Saving Projects with meaningful names (→Project Store)
4. Copying this project multiple times (→Project Copy).
5. Using Project copies for Recordings.

Tempo

Tempo and Tempo changes affect the following operations in the VS:

- →Measure Display (p. 119)
- →Metronome (p. 120)
- →Rhythm computer (p. 194) (VS-2000)
- →Phrase Sequence (p. 155) (VS-2480)
- →Synchronization (p. 217) when the VS transmits tempo data via Midi Clock.

Recorded Audio Data, however, stays the same. The Tempo of a project can be provided in the VS with the following methods:

1. →Tempo Map
2. →Sync Track
3. As a Slave in Midi Clock Synchronization

Application:
Musical →Editing in measures, quarter notes, sixteenth notes etc. makes easy operation in the rhythmical →Grid possible and is still possible even though Tempo changes in the project may have occurred.

Tip:
The function *"Time → Sync Trk"* can be used as a practical calculator to ascertain the tempo of the song and the play time of a passage (→Sync Track).

Tempo Map

The T. serves the musical Tempo and beat settings for a project. This can affect the internal measure indicator, the →Metronome and the rhythm computer of the VS-2000. The Audio Data, however, remains untouched. When →Synchronization is achieved with an external device via →MIDI Clock/SPP, this Tempo information is transferred. The Tempo Map is called up either in the Utility Menu (VS-2000 via the Rhythm Track) or as follows:

Menu: Utility → Tempo Map
VS: SHIFT + TAP MARKER
VGA: SHIFT + Click on TAP

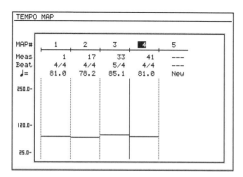

Fig.: Tempo Map

In total, 50 maps can be created. In the diagram above, the T. has already been edited and contains therefore the following maps: Tempos and Time Signatures ("Beat") for the bar numbers 1, 17, 33 and 41.

Map	Bar Nr.	Time Signature	BPM
1	01 - 17	4/4	81,0
2	17 - 33	4/4	78,2
3	33 - 11	5/4	85,1
4	41 - End	4/4	81,0

By default, a Map with a Tempo of ♩ = 120 (→BPM) in 4/4 time is already available. For a continuous Song Tempo and a Project without beat changes, excepting the Tempo setting, no further entries need to take place.

Third Octave Band

Setting the Tempo (♩ =)
The tempo can be changed by moving the horizontal Map line with the mouse or in the row "♩ =" by clicking on the Tempo value and correspondingly dragging. The Dial can also be used. When using the Dial, simultaneously holding SHIFT produces a whole-numbered value change.

Tempo and Beat Changes:
1. Creating a new Map with NEW (F2).
2. In the row "Meas" the bar number for the new Tempo and/or the Time Signature (default setting is measure 2).
3. Entering the new Tempo in the row "♩ =" or the new Time in the "beat" row.

Inserting and Deleting Maps
Via Delete (F4) or Insert (F3), entries can be deleted of inserted. In total, 50 Maps are available.

Tip:
To delete a great number of Tempo Maps, the function "*Time → Sync Trk*" can be used. Under the condition that as yet no →Sync Track has been imported.

> Tempo Map → TM Cnv (F1) →
> Convert Type = "*Time→Sync Trk*" →
> ENTER → YES → YES

The entries can only be changed in intervals of whole measures. For a higher resolution →Sync Track can be used, which imports Tempo information from a Song via →MIDI Sequencer. The difference between the Tempo Map and the Sync Track can be seen in the following table:

	Sync Track	Tempo Map
Resolution	1/96	1 bar
Creation	Ext. Sequencer	VS
Editing	no	yes

A Tempo Map can be converted from a Sync Track (→Sync Track Convert). Tempo editing is also subject to limitations of the measure grid.

Third Octave Band
A →Graphic Equalizer divides the sound spectrum into equally sized frequency bands and levels these then individually. A T. possesses the width of 1/3 of an Octave which in turn produces 31 Bands. The →Analyzer in the VS-24xx models works in the same way with third octave bands. The Graphic Equalizer of the →Effect Patch P031 provides 10 bands in →Octave width.

Threshold
In a →Dynamics processor, when the input signal exceeds or falls below the Threshold level, a certain operation will triggered. Together with the →Ratio parameter, T. determines the amount of lowering/boosting the signal's level in a →Compressor/→Expander. The level reduction here shouldn't affect the entire audio signal but rather only portions above or below a defined level. In this way with a Compressor, Signal portions below this level stay unaffected whilst exceeding this threshold level triggers the level reduction.
→Phrase Divide generates pauses with "Digital Zero" when the level falls below T.

Tick
In a →Sequencer a T. represents the smallest step increment for time positioning of an Event within the predetermined Tempo. It specifies the rhythmical resolution whereby higher values are correspondingly better judged. The VS has a resolution of 1920 Ticks per bar or 480 per quarter note. With a Tempo of ♩ = 120 a tick refers to approx. 3 Subframes (with a →Frame Rate of 30 FPS). →Subframe offers a higher time resolution than Ticks.

Tie
Describes the correction of the note length in the →Phrase Sequence of the VS-2480 and the →Harmony sequencer of the VS-2000.

Time Display
In T. the current →Time Position of the →Timeline (position line) can be read and changed. In

addition to the absolute →SMPTE time in hours:Minutes:Seconds:Frames:Subframes, musical subdivisions, of Measure-Beat-→Tick and markers are available depending on the chosen Tempo.

`00h02m18s29f53  MEAS BEAT TICK  MARKER`
`               046-04-172  006 [Bridge`

Fig.: Time Display in the VGA

In Slave mode, the display can either be shown corresponding to the received Timecode (absolute) or under consideration of an offset (relative). The first beat can be set to any project time (→Sync Offset Time).

Positioning
→Positioning in the project can be easily carried out using the Time Display. To do this, the cursor (a red line in this case) is set via ◀▶ or by clicking on the corresponding time unit and using the dial or dragging with the mouse to enter the desired value. The project is then moved correspondingly over the positions line. Audio material, which has been recorded in using the →Metronome in the project Tempo, can also be positioned according to the time in Measure, Beats or Ticks.

Marker
Setting the current time position to the next or previous →Marker is possible in the Marker field by dragging the mouse or turning the dial. Not only the marker number but also the marker time (without an indication of →Subframes) or Marker Text is displayed.

Fig.: Marker field in the time display

Time Display Format
If an →Offset is set by operating the VS as a →Synchronization's Slave, the →Time Display can be shown either corresponding to the received timecode (absolute) or with consideration for the offset (relative). This occurs for reasons of clarity when viewing both the →SMPTE time and the measure indicator.

Utility → Project Parameter*:
Time Display Format

* *VS-2000: «Display Parameter» menu.*

Time Position
The →Current Time of the →Timeline is shown in the →Time Display as →SMPTE time or in musical measures/beats.

Time Stretching
→Region Compression/Expansion (p. 185)

Timecode
This is an abbreviation for →SMPTE Timecode, which is used for the Tape Recorder and →Video Recorder Synchronization. The SMPTE Timecode is available in the form of LTC (Longitudinal Time Code), VITC (Vertical Interval Time Code) and →MTC (Midi Time Code), which is to be used for VS synchronization. Make sure that the →Frame Rate of the Timecode from →Master and →Slave correspond.

Timeline
The current time is displayed numerically in the →Time Display and graphically in the →Playlist through a vertical red line (otherwise called the "Time line" or "Position line").

TO
The time position TO specifies the target position for the movement of passages (→Edit Points) within →Editing.

TOC
"Table of Contents" refers to the contents of an →Audio CD and a Harddisk Partition. It contains the number of Tracks or data, their order their respective play times etc.

Top
This short cut enables the first passage recorded on active V-Tracks to be located:

Total Recall

VGA:	Hold SHIFT + click ◄◄
VS-24xx:	SHIFT + PROJECT TOP
VS-2400:	SHIFT + REW

Total Recall
Exactly recalling all parameter settings of a →Mixer makes the simultaneous work on multiple projects possible. This option is offered only digitally or by digitally controlled Mixers. In the VS, →Scenes allow a complete programming of all Mixer, Effect and Track parameters, which occurs if they are called up when the device is stopped. Within →Automix, however, the less flexible "Snapshots" can also be used in playback. For storing and recalling of dynamic parameter changes, in the VS the dynamic Automix is used.

TR F/P
The VGA's "Track Mixer Fader/Panorama" page displays the Track and →Effect Return channels where parameters such as →Dynamics, →EQ, →Solo, →Mute, and →Automix can be activated, the →AUX Sends for all Effects can be leveled, the →Effect Patch can be chosen etc.

Home ➔ Page 2/3 ➔ F3 TR F/P

Fig.: Track-Mixer's Fader/Pan Page (VGA, cutout)

Clicking in the FX Return Mixer on an AUX field opens a Flip menu for the assignment of an →AUX Bus to the respective Effect processor. Furthermore, the page TR F/P serves to monitor the →Automix particularly by the VS-2000, which doesn't provide Motor Faders.

TR Mix
To call up the respective →Level Meter, the →Function Buttons can also be used. T. represents here the →Track Mixer's level meter.

PAGE1/3 ➔ F3 (TR Mix)

Track
A Track is a part of the hard disk recorder and contains audio signals or relates to audio signals (→Automix-Edit, →Phrase Sequence). Whilst the VS-24xx models provide 24 Tracks for simultaneous playback, the VS-2000 only has 18 available. A track possesses an additional 15 →V-Tracks, which can be alternatively chosen. These offer a larger variety for practical work. Audio signals of a Track exist in the form of →Phrases and are represented in the →Playlist by a display in the form of a bar. In the VS, Tracks are permanently assigned to the →Track Mixer's channels with the same numbers and can be processed here (→Mixer). For →Editing of Phrases or for →Region Edit, the corresponding Track and mostly also the target track are to be specified (→Track Select).

Track at Once
For →CD Burning of an Audio CD-R, a comfortable procedure with TAO is available, which sequentially creates songs without any time limit. It allows songs to be burned to an already recorded CD-R so long as the operation →Finalize has not yet been carried out and enough memory space is available or the maximum song number of 99 hasn't been exceeded. By doing this, songs can be burned again and again onto a CD without any pressure of time. For the time being, playback is only possible in the VS burner (→CD Player). The CD-R is only compatible with conventional CD Players after →Finalize has been carried out. Multiple songs can also be burned simulta-

neously with TAO whereby the laser according to definition sets a new CD Marker before each new title and generates a pause of 2 Seconds. The created CD-R doesn't satisfy the →Red Book Standard, which can only be realized with the →Disc at Once procedure.

CD-R Write → Write Methods

Track Direct Out
To simultaneously transfer Track signals for editing to external Audio systems, the analog and digital output jacks can be used via of →Direct Outs. The signals are then played behind the →Dynamics and the →Equalizer block. In this operation mode, which is to be activated with T., the Master, Monitor and Aux outputs are not available.

EZ Routing → Output → Track Direct Out

In the default settings, Tracks are assigned to the analog output jacks of the same number. The assignment, as is described in →EZ Routing, can be easily changed. In the diagram, the →R-Bus/Track Routing has been changed.

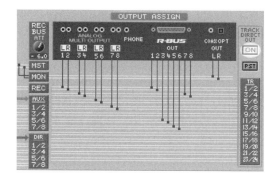

Fig.: Direct Outs (VS-2400)

The signal flow goes from right (TR) up into the physical Outputs via the Busses that are now being used as a Direct Out. The nodes between the horizontally running Busses and the vertically displayed Outs provide information about the current connection. With "Track Direct" activated, the output routing is changed as follows:

Track Exchange

Track	1/2	3/4	5/6	7/8
Analog Out	1/2	3/4	5/6	7/8
Track	9/10	11/12	13/14	15/16
R-Bus Out	1/2	3/4	5/6	7/8

Hint
- When the Direct Out mode is activated, the Aux and Direct Busses are no longer available.
- To listen to the Mix, an Output is to be patched to the monitor bus.

The Tracks can be played either before or after the Fader (→Pre/Post).

Tip
- Single outputs can also be achieved by deactivating "Track Direct out" via →Direct Path and →AUX Busses.
- For editing in external audio sequencers, Tracks/Phrases can be exported to a CD-R using →Track Export/→Phrase Export.

Track Exchange
Exchanging two Tracks together with their respective Track names is realized with the E. operation. Through this, Tracks can be arranged (e.g. all choir voices can be positioned next to each other) and whole Tracks can be deleted.

Hint:
The function can be recalled by using →Undo. At the time of printing, however, this doesn't apply to the accompanying Track names.

"Exchange" can be called up in the Track menu:

Track → Track Exchange

A track is firstly to be activated here and in the same row, the Track that is being exchanged with it should be entered. This function can also be carried out using the (blinking) buttons on

Track Export

the VS via →Track Select (p. 229). This function is nor available in the VS-2000.

Tip
An entire Track is more easily deleted using the "Exchange" operation than →Erase because creating a Region is no longer necessary. An empty Track can be used as the Track being exchanged.

Track Export

Converting entire Tracks into Wave Format and burning a CD-R is called "Export". In this way, recorded Tracks in the VS can be further edited and mixing/mastering can take place in another Audio Systems. Aside from →Sample Rate and →Bit Rate, the time positioning of Phrases and Tracks are also maintained. Blanks before and after Phrases are created by corresponding Passages with "digital 0" in Wave File. Because particularly a single Phrase at the end of a Track can unnecessarily enlarge the Wave File, the →Phrase Export operation is recommended here. Insert a blank CD-R in the tray and operate as follows:

CD-RW/MASTERING → Track Export

The Tracks that are to be exported are indicated with F4 (Mark). Only recorded Tracks possess a check box here so that accidental operation shouldn't occur. The list in the "Map" Menu, however, shows all the →V-Tracks more clearly. As the next step, the CD menu is called up using F5 (Next). The Parameters "Write speed" and "Verify" (set to "On") need to be set. In the TRACK field, the selected tracks are listed for double checking. The required memory space is shown under the list. E.g. "5 Track 1205, 35MB" will require two CD-Rs. F5 (OK) starts the burning of the Wave Files to the CD-R.

Tip:
Single →Phrases can be burned to a CD using the →Phrase Export operation.

Track Import

Using Tracks from other projects in the current project is realized using the "Import" operation. T. represents a physical copy of the source Track in the current project and therewith enlarges the corresponding memory on the harddisk partition that is needed.

Application:
- Using Drum tracks from other projects multiple times.
- Copying →Master Tracks of other projects into the existing project to create a CD.

The →Record Mode (p. 175) and the →Sample Rate (p. 199) of both projects must correspond in order for this to work. These must also both be in the same →Partition. Only one single Mono Track can be imported which means that the operations for L and R of the Stereo Mastering Tracks must be carried out twice.

Track → Track Import

The "Source Project" Menu is similar to the →Project Select one. Firstly here, the Source project, which contains the desired Track, is to be marked using F5 (Select). The detailed V-Track to be copied in the "Source" field is then to be selected. The target Track for the copy is then chosen under "Destination" for the current project. After several prompts, which are to be confirmed, the VS copies the Source Track to the current Project.

Tip:
For Target tracks that already contain phrases, the Source Track can be added after the last phrase with a pause of 2 seconds. The following then appears: *"Import Source Track at the back of Destination?"* This option allows easy consecutive copying of Master Tracks for creating a CD, which can be created in different projects.

Track Mixer

Among the →Mixers in the VS, the T. serves to enable listening and editing of the Track Signal. Signal editing carried out here displays purely playback parameters and can always be returned to.

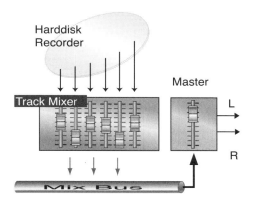

Fig.: Track Mixer

The Fader unit is to be firstly assigned in the Track Mixer for editing via the →Fader Buttons In the VS-2000, the Faders are hard wired to the Track Mixer's Channels.

Track Name

All →V-Tracks of a VS project can be provided with names. Particularly when →Mixing but also when editing Tracks, this provides considerable relief. Entering the →Name Entry via a PC keyboard also makes it easier.

Track → Track Name → select → Name

Track Phrase Edit Menu

Aside from Mouse operations, all Phrase operations (→Phrase Edit) in the LCD can be carried out in this Menu. Calling this up is achieved via the global Track Menu. Firstly, select the Phrase Edit Menu via F6 (→Function Buttons). This button toggles between the Phrase and the →Track Region Edit Menu. In the VGA, all operations are combined in the Track Menu.

Track Region Edit Menu

In the LCD, the T. contains both the Operations for →Region Edit and the functions purely relating to Tracks (→Track Name, →Track Import and →Track Exchange). In the VGA, all operations are combined in the Track Menu.

Track Select

The first step for Track/Region or Phrase Edit operations that are to take place is to select one or more Tracks.

Playlist
In the playlist, position the Track Cursor with the cursor button ⇕ over the desired Track. This is displayed inverted and then selected.

Edit Menu
In the Edit Menus, F1 (SelTrk) calls up the Select page. Clicking in the field on the left before a Track marks the current V-Track. Alternatively F3 →Mark can be used. If another V-Track should be activated, this can be carried out using the Dial or the Mouse. The F2 field serves to select all active Tracks (All).

Buttons on the VS
Excepting the VS-2000, within certain Edit Operations, the required Track assignments can be carried out directly using the VS panel's buttons (for active V-Tracks). Here, the CH EDIT button acts as the source and the TRACK STATUS button, as the target tracks. Blinking buttons represent unassigned Tracks.

Track Status

T. describes the operation status →Play, Record, →Record Monitor and Off of a Track. It is irrelevant here if this represents a recorded V-Track or not.

VS Panel
The desired T. is achieved by pressing the respective STATUS button multiple times. By holding Play and activating a STATUS button or by holding REC, respectively Playback and the record status can be directly chosen.

VGA
- In the Playlist, selection takes place by clicking on a status field (PLY by default). All tracks remain to PLAY via the yellow arrow.

Transport Buttons

- In the pages →TR F/P and →MltChV, the status fields are located above the faders.
- In a single Track-Mixer's →CH EDIT page, the Status is to be found below the CH No.
- Using →Control Changes, the status can be externally controlled by a sequencer or another VS (→Synchronization).

Transport Buttons

For simple positioning within the project, the well known Transport functions are available:

Button	Function
PLAY/STOP	Playback/Stop
ZERO	To the Song's start
REC	Recording
ZERO	To 00:00:00:00:00
REW	Rew or 1s back
SHIFT + REW	To the 1st Phrase's start
SHIFT + FF	To the last Phrase's end
FF	FF or 1s forward

In the VGA, clicking on the corresponding Transport Icons leads to the same result. Additionally it is possible to move within the project with the operations described in →Positioning (p. 162).

Trim In/Out

Using Phrase TRIM the Start and End of a Phrase can be shortened (→Phrase Trim). This Operation is generally used for cleaning up a phrase.

True Stereo

Although common Reverb processors provide inputs for R and L, generally both channels are mixed together and a Mono signal will then be processed. Only "True Stereo" devices process both channels independently and also maintain the Stereo Positioning in the Effect. In the VS, →SRV Stereo Reverb and →TCR 3000 Reverb represent purely Stereo Reverbs.

Tuner

Only the VS-2000 provides an integrated, chromatic tuning device, which can be routed to any Input or Track Mixer channel for checking the pitch.

VGA:	Click on the TUNER Field
VS-2000:	TUNER button

Assignment firstly takes place in the source Field of the desired Mixer Channel. After choosing the reference tuning, which normally averages 440 Hz, the indicator is achieved. An exactly vertical pointer position refers to a correct tuning.

THRU
Activating F1 (Thru) routes the channel signal to the →Mix Bus whereby it can be listened to regularly (See Tip).

GET
Tunings that differ from the norm can be used as 0- reference if they are a maximum of 1/4 tones above or below 440 Hz. In order to do this, enter the corresponding Tone and press GET. Furthermore, the Indicator adopts the 0 Position and then acts as the new reference tuning. This is shown in the Pitch field.

Tip
The Tuner button, diverted from its intended use, can operate as a mute switch for any Mixer Channel. In this way for example, a →Talkback microphone can be turned on and off (deactivate THRU).

U

Unbalanced

In contrast to a →Balanced connection, U. provides only one signal wire inside a ground shield. The shield offers sufficient protection from low-level interference and constitutes the 0V potency. The transmission of high leveled signals with →Line Level (CD Players, synthesizers, etc.) is unproblematic with →Cables of moderate length (up to approx. 5m). Digital connections used in the VS (→S/PDIF, coaxial) also exhibit an asymmetrical connection.

Undo

Due to →Non-Destructive Editing (p. 135), editing of Audio Data taking place in the VS is not carried out in the original →Take (p. 221)s but rather in their virtual copies. Operations can be therefore canceled without accessing the original. The VS Undo affects audio recordings, →Region Edit and →Phrase Edit whilst Mixer and System settings are not reset. In total, 999 Undo Levels are available, which makes canceling up to 999 editing steps possible.

VGA:	Click ▭▭ (Undo/Redo)
Mouse:	Rightclick in the playlist
VS:	UNDO button

All Operations of the current project can be itemized with Names and Date/Time in the called up list. These can be accessed to view their current status (Levels). Because the last operation is generally reset, merely press YES here. Editing steps further back than the last one can be called up by selecting the corresponding list entry. It is important to remember that by going back to previous editing steps, all following operations are also cancelled (see Tip).

* To save time, the last editing can also be carried out without the →UNDO Message.

Redo

The last Undo procedure carried out can also be cancelled if the project hasn't been saved yet. This returns the project to its status previous to the last Undo Operation and accesses all Undo Levels in the list:

VGA:	Click ▭ (Redo)
Mouse:	Right click in the playlist
VS:	SHIFT + UNDO

Note

After →Project Optimize has taken place, all Undo Levels are erased. Undo and Redo can then no longer be carried out.

Tip

Whilst with U. the entire project status with *all* recording and editing steps is restored, →Phrase New allows original Phrases to be restored in the current stage of the project.

Special Undo functions with only one Level are additionally available in →Automix and in the →Phrase Sequence(VS-2480).

UNDO Message

Generally the →Undo operation is activated via the Undo table, where the selection by name/date of creation can be comfortably carried out. For a faster approach this table can be discounted and the last operation can simply be recalled without any prompt. This setting is only recommended for pro users:

VS-24xx: Utility → Global2 → Undo Message
VS-2000: Utility → System Prm2 → Undo Message

Universal Audio

The company U. adapted their →Plug-ins →LA-2A (p. 102) and →Urei 1176 Compressor (p. 231) for the Roland platform VS8F-3.

Update

→System Update (p. 219)

Urei 1176 Compressor

This →Plug-in from the company "Universal Audio" emulates the vintage →Compressor and Peak limiter "Model 1167LN", which is distinguished by "analog warmth" and extremely short attack times.

Fig.: Urei 1176 LN Compressor Plug-In

USB

It is suited to be used for the Compressing/Limiting of Drum and Guitar sounds. Using the legendary "All Button Mode", (pressing all ratio buttons simultaneously), an extreme compression is achieved for special sounds.

Urei 1176 LN

Effect Patch:	Plug-In
Connection:	Insert
Application:	Drums, special

Operation:
The compressor is switched on by selecting one of the a Level Meter Modes (GR, +8, +4dB). Using the METER button, either the output level or the gain reduction indication is chosen. In comparison, the setting OFF activates a Bypass. Firstly, with the OUTPUT knob, set to "-∞", choose one of the pre-assigned Ratio values and set METER to GAIN. Leveling the input signal with the INPUT knobs here simultaneously specifies the Threshold Value. After the desired Gain Reduction is achieved, the necessary output volume can be leveled via OUTPUT. Set the applicable Attack and Release Times depending on the signal to be processed. The legendary "All Button Mode" is realized using F4 All.

USB

USB2.0 represents a standardized interface for exchanging data between computers and other devices with a transfer rate up to 480 Mbp/s. Amongst the VS models, only the VS-2000 allows Audio and Project data to be transferred to a computer and back again. Furthermore, →System Update is also possible here using USB.

USB-Connection
1. Establishing the Connection
Activate the USB buttons on the VS or on the corresponding buttons in the main VGA page.
After various security prompts, the mode "USB2.0" is activated. In this mode of operation, the VS cannot be operated. Subsequently, the VS and the computer are connected with a conventional USB cable. The message "Connected" is then displayed. The computer's desktop displays the additional four VS-2000 partitions.

Fig.: VS-2000 Partitions in Mac OS X

2. Terminating the Connection
To terminate the connection, the VS drives must first be dismounted on the Computer Desktop, before the USB connection can be terminated using the VS's EXIT. The USB cable can then be disconnected.

Wave Converter
With the help of the "Wave Converter" program of the enclosed CD Rom, Tracks can be transferred from the VS-2000 to the computer and vice versa. On the one page of this program the partition can be selected under "VS-2000 Drive" and the desired project under "Project". In the V-Track Table recorded Tracks are displayed in yellow.

• Transferring VS-Tracks to a Computer
By clicking on a recorded V-Track (yellow) the color changes to orange. Using «VS Track → Wave File» the window for Name entry and specifying the path of the target folder in the computer is opened. "Save" converts the Track and saves it in the indicated folder. Creating a stereo file by selecting any two V-Tracks is also possible.

• Transferring a PC-File to the VS
By clicking on an unrecorded V-Track, its color changes from grey to red. A file can be chosen with «Wave File → VS Track» and exported to the indicated V-Track using "Open".

Use Effect Board
The →Harmony function of the VS-2000 requires the Effect board →VS8F-2 for the generation of harmony voices, which are to be assigned using this setting.

User
For →Effects, →EZ Routings and the VS-2000's →Rhythm Track, factory presets are available. These cannot be changed. One's own settings can be saved in the User Memory locations. In the default settings, the User memory locations generally contain copies of the Presets. They can therefore be overwritten without another thought.

UTILTY Menu
The U. allows access to the following pages. The basic VS parameters are set here:

→System Parameter (p. 219)
→Global Parameter (p. 85)
→Project Parameters (p. 169)
→Play/Rec Parameter (p. 158)
→MIDI (p. 122)
→Sync Parameter (p. 215)
→Tempo Map (p. 223)
→Metronome (p. 120)
→Auto Punch/Loop (p. 18)
→Marker (p. 109)
→Locator (p. 106)
→Scene (p. 200)
→V.Fader (p. 235)
→Surround (p. 213)
→Generator/Oscillator (p. 85)
→Analyzer (p. 16)
→R-Bus (p. 172)
→Control Surface (p. 49) VS-2480
→Automix (p. 19)
Phrase Sequence →Sampler (p. 199) VS-2480 only
→Date/Time (p. 52)
→Parameter Initialize (p. 142)

To create a →New Project, the Utility settings of the current project can be adopted (→Copy Utility Parameter (p. 50)).

V

V-Fire
The optional Audio interface from the company PreSonus allows audio data to be exchanged between a VS-24xx and a computer via Fire wire. It provides two →R-Bus connections and can therefore simultaneously transfer 16 channels of a VS-2480 in both directions with a maximum →Sample Rate of 48 KHz. Because the PCI card →RPC-1 from the company M-Audio is no longer available, the V. is the only* possibility for digital transfers to a PC.

By using the AES/EBU interface →AE-7000, eight channels can be transferred via R Bus.

Fig.: PreSonus's R-Bus to Firewire Interface

V-Link
For controlling Video editing devices of Roland's subsidiary company "Edirol", standardized Midi parameters can be used. By activating the V-LINK button (not on the VS-2480) both devices synchronize automatically and allow Image effects, fade overs, Slow/fast motion etc. to be carried out with the VS Faders.

V-Track
Aside from the currently selected (main) →Tracks, additional (minor) tracks otherwise known as V-Tracks are available in the VS. Whilst the number of Tracks that can be simultaneously played is determined by the system to 24 or 18 in the VS-2000, the V-Tracks represent a Track reserve from which respectively one Track is chosen to be played back. Each Track, therefore, provides 15 V-Tracks in addi-

V-Track

tion to the current Tracks that can be played back. These have principally the same worth and are merely inactive at the time. It can be inferred from this that by selecting another V-Track, all current (main) tracks can change to an inaudible V-Track. In this way, recordings can only be carried out in selected V-Tracks, whilst copies are also possible on non selected V-Tracks. By calling up a V-Track, the →Mixer settings of the corresponding Mixer channel are adopted. In particular, these are the volume, →EQ, →Dynamics, assignment of →AUX Busses, →Effects etc. The VS models provide the following number of V-Tracks:

  VS-24xx: 384 V-Tracks
  VS-2000: 288 V-Tracks

The relationship of a V-Track to a Main Track can be derived from its labeling. In this way, "3-14" or "V.T3-14" refers to V-Track 14 of Track 3. To differentiate, stereo coupled Tracks (→Link) respectively have an additional "R" or "L": e.g. 15L-1 and 16R-1 are available for the first V-Track of the Stereo Track 15/16.

Current Track	V-Track
Selected / Playable	Not selected/playable
displayed	not displayed
recording possible	no recording possible
graphical/numerical Editing	only numerical Editing

Copying Passages is possible not only within a V-Track of a Track but also comprehensively of other Tracks. E.g. 3-15 → 16-11.

Application

- Alternatively instruments being played in or different vocal recordings can still be chosen in the →Mix (p. 125).
- By →Bouncing →Effect Returns to different V-Tracks of a Track, previously recorded Effects can be chosen (→Effect Economy (p. 65)).
- Mix varieties by mixing to different →Master Tracks (p. 118).

Selection

A V-Track is selected only whilst the project is stopped. Because principally all V-Tracks can be provided with a name (→Name Entry), this name appears in the playlist when the V-Track is chosen as an active (main) Track. The assignments of V-Tracks are saved in →Scenes and can be easily managed in this way. The following methods are available for selection:

1. Track Mixer Channel

 By pressing one of the →CH EDIT buttons on the VS or in the VGA by double clicking on the channel number, the corresponding →Channel Strip of a Track opens in the →Track Mixer. A V-Track is selected here in the «V.Track» field by clicking and dragging or by turning the dial. In the LCD, the Flip Menu only opens after being positioned over the field and confirmed with YES.

2. VGA playlist

 The second column in the →Playlist indicates the particular V-Tracks. Clicking on a Field opens the Flip Menu for V-Track selection.

3. V-Track-Map

A complete list of all V-Tracks of the current project occurs in the V-Track Map.

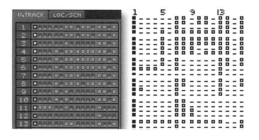

Fig.: V-Track Map in the VGA and LCD

This is shown in the same way in the VGA and LCD and makes an overview in the table of recorded/unrecorded and selected/unselected V-Tracks possible (→Tab.: Symbols in the V-Track Map (p. 235)).

V.Position

3.1. Selection in the VGA Display

 Clicking on ▸ in the VGA Display opens the control field for the V-Tracks, Locators and Scenes. This window can also be directly called up on the VS using SHIFT + HOME: Within the Table, a yellow, filled in square refers to a recorded, active V-Track.

3.2. Selection in the LCD
The V-Track Map in the LCD of the VS is present in the →HOME page. It can be hidden in order to enlarge the playlist:

SHIFT + HOME (several times)

When the LCD is used as →Operation Display, a corresponding Mouse click here selects the desired V-Track. Within the table, a filled in square refers to a recorded, active V-Track.

V-Track	LCD	VGA-Display
Recorded, Selected	Box Filled	Blue Box Blue
Recorded, unselected	Box Empty	Yellow Box Empty
unrecorded, selected	Half Box Filled	Half Box Yellow
unrecorded, unselected	Line	Line

Tab.: Symbols in the V-Track Map

All V-Tracks
In the editing operations →Region Erase, →Region Insert and →Region Cut, it is possible to expand editing to all marked tracks of the 16 →V-Tracks using "AllVTr".

Application
- Simultaneously →Eraseing passages of all 16 V-Tracks of one or more Tracks for global changes of an arrangement. This is important for alternative varieties on the V-Tracks of a Track (e.g. Different Singers were recorded, which were to be chosen from at a later date).
- Changes in the arrangement of all V-Tracks.

V.Fader
Aside from their main function as a volume controller, the Faders can also be used as Midi Controllers (not in the VS-2000). By sending defined →Control Changes, the corresponding parameters can be controlled in Midi Sound modules or →Sequencers:

VGA:	Page 2/3 → F4 (MltChV) → V.Fader
VS-2480	SHIFT + V.FADER
VS-2400:	V.FADER

Application:
- Volume, Panorama and Effect control of Midi instruments of Midi Sound Modules.
- Midi Faderbox
- Use as a →Control Surface for the Midi/Audio sequencer.

In the default settings, the Faders send CC 7 (Volume) via the corresponding Midi Channels and additionally CC10 (panorama) to the Pan knob of the VS-2480. Individual assignment is created in the V.Fader page of the VGA display or with Utility → V.Fader in the LCD.

Note
The →Fader Assign or →Knob/Fader Assign parameters in the VS-2480 provide the assignment of an internal mixer parameter to the Faders.

VS-2480: →Control Surface Templates for conventional Midi/Audio sequencers are available for downloading on the Roland Website.

V.Position
In the →VGA Monitor, the V. serves to adjust the display vertically with a setting range of -5 to +5.

Utility → System Parameter2 → V.Position

Value

Value
An abbreviation for "Parameter Value" in →Parameter Edit (p. 142).

Vary Pitch
Changing the record/Playback speed of a Recorder, with the corresponding change of pitch is referred to as V. It especially serves to adjust a recording so that it matches the recording of an instrument that can't be subsequently tuned such as an acoustic piano, accordion, organ etc. It also serves to help singers when a passage's pitch is too high/low. The V. function can also be used for alienation purposes (→Mickey Mouse Effect). Whilst in analog Tape Recorders the capstans rotation speed is varied, in hard disk Recorders the →Sample Rate is increased/decreased.

Utility → Play / Record1 → Vari Pitch.

Interval	Lower	Higher
Prime	44,10	44,10
Minor Second	41,66	46,72
Major Second	39,30	49,60*
Minor Third	37,09	-
Major Third	35,0	-
Fourth	33,06	-
dim. Fifth	31,21	-
Fifth	29,44	-
Min. Sixth	27,77	-
Major Sixth	26,23	-
Min. Seventh	24,75	-
Maj. Seventh	23,36	-
Octave	22,05	-

2 Cent too high

The table shows the changes to the pitch in relation to the Sample Rate based on 44,1 KHz (prime) in KHz. In the VS (not the VS-2000), transposing up to one octave lower and a second higher is possible.

Note
For the →Recording of digital Signals and →CD Burning, the Sample Rate cannot be varied.

Tip:
An external Clock, which clocks in the VS via →S/PDIF, determines the pitch of Tracks, which have already been recorded. This is the same as a Vary Pitch.

VE-7000
This external →Channel Strip provides dedicated Encoders for all parameters of the →CH EDIT menu the parameter control is carried out via →MIDI. Aside from easy use, it is suited for →Surround mixing particularly because of its integrated Joystick.

Fig.: External Channel Strip VE-7000

After calling up a Mixer Channel with the CH EDIT button, the Channel Strip is assigned and allows Parameter changes for the channel, which has been selected in this manner.

VGA Monitor
Using an optional VGA Monitor combines the advantage of graphical visualization of Computer supported systems with the operation safety and ergonomic usability of pure Hardware →DAW. A conventional PC monitor (XGA, resolution 1024 x 768) is used here, which is to be connected with the DB 15 Type connector of the VS. In the VS-2000, the optional Board VS20-VGA is required for connecting the monitor and the mouse.

Note
For mouse editing in external VGA monitors, they are to be specified as →Operation Display.

The following Parameters serve to activate and adjust the external monitor to the VS:

    Utility → System2 → VGA-Out
                                     Refresh Rate
                                     H./V. Position

Refresh Rate:
How often the display is refreshed per second is determined by the parameter R. For a TFT display, the slowness determined by the system is able to be selected in rates of 66-75 Hz, This selection can be made depending on the display quality. Generally and particularly for tube monitors, this setting should be carried out with the dial in order to be able to return to the last Value in case of picture failure.

H. and V Position:
These Settings serve to adjust the picture position both in vertical and horizontal directions. Generally the automatic function of the monitor centers the picture correctly.

VGA Out
This Parameter is used for activating an optional →VGA Monitor:

        Utility → System2 → VGA-Out

Video Recorder Synchronization
Digital and analog video recorders can be used with the applicable →Synchronizer for →Synchronization as the Master for the VS. By doing this →MTC and →MMC will always be generated from the received →Timecode, which externally controls the VS. The Midi and Sync pages for setting the necessary Parameters are called up in the Utility Menu:

Midi-Parameter	Sync-Parameter
SysEx.Rx Sw = On	Sync Mode = Ext
MMC Mode = Slave	Frame Rate like Master

DV-Video Recorder and Camcorder
Due to the LANC or EDIT E connection of popular DV Video Recorders and Camcorders, the digital Video Timecode can be converted to MTC/MMC using the optional →SI-80SP synchronizer. In doing this, the VS is controlled as a slave exactly in frames not only in playback, but also in FF/REW and Slow motion. For European video productions, a value of 25 Frames/s is to be chosen as the →Frame Rate.

Analog Video Recorders
An Analog VTR as a Master can also synchronize the VS. To do this an additional synchronizer needs to be used, which converts →SMPTE to →MTC. Firstly the synchronizer's Generator function is used to record the SMPTE Timecode onto the VTR's audio track. When dubbing to a further VTR on one of the two Hi-fi Tracks, the movie sound is then to be recorded to the remaining track. While Playing back, the Timecode is fed into the Synchronizer which converts the SMPTE into MTC. Finally the MTC/MMC controls the VS as described in the synchronization chapter (→2. VS as Slave (p. 218)). In contrast to DV Recorders, the Timecode is only transmitted during playback. When in slow motion or moving in FF/REW, the VS correspondingly stops.

Vocal Cancel
To eliminate signals positioned in the stereo center of an entire mix, the →Effect V. can be used. Because of the localization of further instruments in the stereo center, Vocal canceling cannot be carried out in isolation. This means that sound interferences are unavoidable.

Vocal Canceler	
Effect patch:	P 024
Connection:	Stereo Insert
Application:	special

Deleting the center signal can be limited to any determined frequency range. In this way the vocal frequency spectrum can be determined between approx. 100 Hz with "Range Low" and 3 KHz with "RangHi". If the vocal signal is not centered, the processing position can be adjusted correspondingly with "Balance". For further frequency corrections a 3-Band →Parametric Equalizer is provided.

Vocal Channel Strip

The V. →Plug-in is included in the delivery of the →VS8F-3 Effect Board. Both independent →Channel Strips additionally provide all effect blocks necessary for vocal editing. Two mono channels or one stereo channel strip can be operated with a single Plug In processor. It can be used with project Sample Rates of up to 96 KHz. Due to the large number of factory presets, a proper patch can be utilized quickly and for any application. Furthermore, the presets allow their typical sound shaping parameters to be studied and adopted by the →CH EDIT Parameters and Effects.

Vocal Channel Strip

Effect patch:	Plug-In
Connection:	Insert 2x Mono or 1 x stereo
Application:	Vocal, Instruments

Firstly the particular Mono effect is to be inserted in the desired mixer channel via the proper →Insert-Mode. In the Plug-In menu, the corresponding page can be called up (R or L)

Example:

Plug-In Block	Mixer channel	Connection
L	Input 3	
R	Track 9	

The →Compressor reproduces sounds both of Solid stage and various Vacuum Tube circuitry and provides a →Soft Knee characteristic if necessary. With the →Enhancer/ →De-Esser, which can be alternatively used, higher frequencies can be accentuated or lowered. A →Parametric Equalizer allows typical frequency corrections whilst →Pitch Shifter, →Chorus and →Delay serve creative sound purposes.

Vocal Multi

V. refers to certain necessary →Effects typical for the editing of vocals within a Multi effect with multiple Blocks available: A →Noise Suppressor suppresses background noises, the subsequent dynamic processor can either be used as a →Limiter or as a →De-Esser, an →Enhancer raises higher frequency rates that are dependent on levels, a three band →Parametric Equalizer serves to align the frequency flow, →Pitch Shifter, →Delay and →Chorus add artistic sound elements.

Vocal Multi

Effect Patch:	P139 - 145
Connection:	Insert
Application:	Vocal

A large number of the editing tools can be alternatively realized via a channel strip's tools and →Send/Return Effects.

Vocoder

The Vocoder →Effect (VOice and enCoder) encodes a signal (mostly spoken) by impressing its dynamic frequency spectrum onto a carrier signal. Through this, the Effect of a "Speaking" instrument is produced. The articulation of another instrument can also be used.

Application

- Speaking Instruments
- Experimental sounds

Active Principle

In the analyze unit of the V., up to 19 →Bandpass filters are available, which subdivide the input vocal signal (modulator) into the predetermined frequency ranges. Control signals are generated from the individual signals of the filter bands and are fed into the synthesis unit. The same Bandpass Filters as are in the Analyze unit are available here again for the carrier signal. These also disperse the carrier signal into frequency ranges whose volume is no longer specified by the carrier signal itself but rather from the synthesis unit's control signals for the corresponding filter bands. The current volume of every modulator's 19 filter bands are therefore transferred to the synthesis signal. The result of this is that preexisting Frequencies that have been boosted in the Modulator, are also outputted more loudly in the corresponding frequency band of the carrier signal. Finally, when all modulated, individual Band pass signals are combined together, they mirror the input signal with the sound of the carrier - the instrument talks.

In the VS, the Instrument is to be fed in via the left →Insert Block of the Effect (synthesis unit) and the voice correspondingly in the right Insert block (Analyzer unit) whereby any Mixer channel can be selected. The basic sound is determined by the number of Band passes and their volume settings, which can be changed using the "Character" parameter. In the VS, two Vocoder Algorithms are available:

[Vocoder]:
The Vocoder1 provides 10 Band passes for simple use with medium quality and only requires the processing power of any one chose Effect processor.

[Vocoder2(19)]:
The qualitative high value 19 band Vocoder2 requires the power of two processors (1,3,5 or 7) due to its high processing demands.

Vocoder

Effect Patch:	P180; 181
Connection:	Insert R + L
Voice:	Ins. right
Instrument:	Ins. left
Application:	Special

Voice Transformer

Because of its independent editing of pitches and Formants, this vocal →Effect generates unusual sounds from a solo (isolated) voice. In this way, male and female voices can be converted, female voices can be created from male voices, children's voices, robot and drone sounds etc. Additionally, the VT* offers its own reverb for spatial effects.

Voice Transformer

Effect Patch:	P105- P109
Connection:	Insert
Application:	experimental

Editing
Results can be easily achieved by experimenting with the chromatic knobs *Pitch* and *Formant*. For a light, natural sound color change, the pitch should be left at the value of 0 and the Formant value should be changed by a maximum of 3 half tones. For a second voice (e.g. octaves) leave Formants at a value of 0. With *Robot* = On, the original voice loses its tonality and the pitch can only be specified then via Pitch. Aside from typical Robot voices (Pitch = -12; Formant = + 12), drone voices (such as the bagpipes) can also be created.

* *The VT is also available as "Boss VT-1".*

Volume
→Level (p. 103)

VS-1680, VS-1880, VS-1824

The previous models of the Series 2 provide 18* play back tracks / 8 Recording tracks and a 10 GB hard drive. The maximum partitions size is 2 MB and the best audio quality can be achieved with the →Record Modes →M16 or →MTP.

Fig.: VS-1824

VS-2000

The VS-1824 is fitted with a CD burner whereas the previous models offer an external SCSI burner. Four Aux busses allow the signals to be outputted to external targets or to the four (optional) effect processors. The channel strips (not motorized faders) provides a simple equalizer, Dynamics can only be used with the help of Effect Inserts. The connection of a VGA monitor is not possible.

VS-1680: 16

VS-2000

The smallest and latest of the Series 2 models, it is not fitted with motor faders. It records eight uncompressed (16 or 24 bit linear) tracks simultaneously with a →Sample Rate of 44,1 KHz. With a →Word Length of 16 bits, it provides 18 playback tracks, 24-bit reduce it to 12. A monitor/mouse connection is optional and requires the board. It provides a graphical display with 320 x 240 Pixels and a size of 10 x 8 cm. The internal rhythm computer, the guitar tuner and the USB connection distinguish this model from all the others of this series.

Fig.: VS-2000

Details can be found in the appendix's →Comparison Chart of Functions (p. 246).

VS-20 VGA

The →The VS-1824 is fitted with a CD burner whereas the previous models offer an external SCSI burner. Four Aux busses allow the signals to be outputted to external targets or to the four (optional) effect processors. The channel strips (not motorized faders) provides a simple equalizer, Dynamics can only be used with the help of Effect Inserts. The connection of a VGA monitor is not possible. requires this optional board for operation with the mouse/VGA. It provides connectors both for the mouse and the optional VGA monitor.

VS-2400

The VS-2400 simultaneously records ten* Tracks and plays back 24 Tracks without any additional Hardware. The →Sample Rates and →Record Modes correspond to those of the top VS-2480 model. It provides 13 Motor faders and a graphical display with 320 x 240 Pixels and a size of 10 x 8 cm.

Fig.: VS-2400CD

Details can be found in the →Comparison Chart of Functions (p. 246) in the appendix.

8x analog, 2 x digital (S/PDIF); with optional →ADA-7000 16 x analog

VS-2480

The top model of the VS-2xxx series provides over 17 motor faders and a large graphic display (12 x 9cm, 320 x 240 Pixels), which is also designed for operation without an additional monitor. Thanks to it's fully equipped number of preamps, 16-Track recording to is provided. The internal Sampler function, SMPT IN, 2 R-Bus connectors and the additional filter in each Mixer Channel are to be distinguished here. Details are to be seen in the →Comparison Chart of Functions (p. 246) in the appendix.

Fig.: VS-2480

VS-880, VS-880EX; VS-890

In 1995 with the production of the VS-880, an economical 8 Track recorder was introduced for the first time.

Fig.: VS-880EX

It records a maximum four tracks simultaneously and provides a total of 8 V Tracks for each Track giving a total of 64 V Tracks. The display is shown in two columns, which demands a certain amount of ability in abstraction from the User. It was not designed with a connection to a VGA monitor.

VS8F-2
→Effect Board (p. 64)

VS8F-3
→Effect Board (p. 64)

VSR-880

The VSR is used especially in conjunction with the Roland digital mixers of the VM series as a Rack Version of the VS-890 due to its R-Bus connections. The maximum →Partition (p. 143) size is 2 MB. It features two mic inputs (without →Phantom Power) and 8 line Ins.

Fig.: VSR-880

In the special VSR operation mode (without EQ) the →Record Mode →MTP is already provided.

Wave
1. wav

W. represents an Audio data format originally created by Microsoft for purely multimedia applications, which because of its widespread distribution almost equates to a standard. Wave Files provide no timecode information and must therefore be positioned manually in the →Playlist after importing them. They can exist as either compressed or linear PCM data. In the VS, Tracks and Phrases can be exported in the Wave format (→Wave Export). Imported Wave data is applied as a new phrase (→Wave Import).

2. →Waveform (p. 242)

WAVE DISP
→Waveform (p. 242)

Wave Export

Wave Export

For further editing in external Audio Sequencers, either entire Tracks can be burned as Wave Files to a CD-R using →Track Export or single phrases can be burned using →Phrase Export. Via a →USB connection, the VS-2000 is able to carry out the transfer directly into the PC/Mac Processor. To do this, use the "Wave Converter" software, included in delivery.

Wave Import

In the VS, Wave Data can be imported in a linear PCM Format and can then be used within the Project as a new →Phrase. The transfer occurs using a CD Rom (ISO 9660). Only the VS-2000 allows additionally allows a →USB Transfer.

CD-RW/Mastering → Wave Import

In the called up list, It is possible to pre-listen to WAV. Data using F2 (Preview). Using F4 (Info) the information about size, play time and production data is called up. F5 (Select) opens the "WAV Import" window:

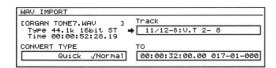

Fig.: Importing Wave-Files

The Target track, the type of conversion for formats that don't correspond to the Project's format ("Normal" provides the best quality), and the time position are chosen here.

Tip

The →Current Time of the Timeline is adopted for TO using →Get Now (F2) and navigates the beginning of the Wave file to this position.

OK starts the operation. Importing requires a certain amount of time. The name of the imported Wave File is adopted as →Phrase Name.

Waveform

The W. option in the VS (Wave) allows a graphic display of an Audio Signal. The volume is represented by the Y axis (amp.) and the time by the X Axis (Time). Automatically scrolling the wave forms is not supported.

The chosen zoom level is also shown alongside the Track number, the name of the track and the current →Phrase.

Fig.: Wave Form Display in the VGA

Application

- Finding →Edit Points within the →Region Edit and →Phrase Edit
- →Clean Up
- Editing in connection with →Scrub

1. Choosing a Passage

The Wave form display is limited to one Track. It is therefore to be selected in the →Playlist using the Cursor ≑ or by clicking. Subsequently, the →Timeline is to be approximately positioned over the passage.

2. Calling up

VGA:	Click on 〰 bottom right in playlist
VS:	SHIFT + F5 (Wave Disp)
VS-2480:	WAVE DISP button

3. Positioning

Within the Wave form display, the visible cutout can be positioned in the →Time Display using the Operation described in →Positioning. Generally "Frames" or "Subframes" are selected with the Cursor button ◀▶ or by clicking. Positioning then takes place by turning the dial or by dragging with the Mouse.

4. Zooming and choosing a Track
4.1 Zooming
Corresponding enlargements can be carried out not only in the X Axis (Time) but also in the Y Axis (Amp.) using numerous methods:

VGA:	Right click in the Wave window →	ZOOM ↑↓ ZOOM ↔
VGA:	Click in "+" or "-" of scroll bars	
VS:	SHIFT + Cursor ◄►, ▲▼	
Keyboard:	SHIFT + ◄►, ▲▼	

4.2. Choosing a Track
It is also possible to select Tracks in the wave display:

VGA:	Vertical →Scrolling
VS:	Cursor ▲▼
Keyboard:	Arrow buttons ▲▼

5. LCD
The Wave form display can also be called up in the LCD. Hiding the Level Meter on the screen achieves a display which fills the entire screen:

SHIFT + HOME

Tip
When in operation with the VGA, the LCD can be used for independent Wave displays (→Information Display). To do this, call up the 3/3 page in the playlist display via PAGE and select F4 (ID Wave). F6 then locks this in against a page change in the →Operation Display.

Wheel/Button Editing
The Time/Value dial and the Cursor buttons offer editing for especially difficult editing procedures. This combines the advantages of clear graphic editing with those of precise numerical editing. A selected Phrase or Region in the Playlist is not selected for this with the mouse but rather graphically positioned with the On Board operating elements of the VS.

1. Erase TO in the Edit Memory location: SHIFT + CLEAR + TO
   (→Edit Points, →Edit Point Buttons, →CLEAR)
1. Select a Phrase or Region: (→Phrase Edit, →Region Edit).
2. Choose the Operation (E.g.: COPY): Right click in the playlist and select COPY (VS-2480: COPY button)
3. Move the copy (displayed inversely) to the desired Position:
   Dial: Time Positioning
   Cursor Up/Down: Track
4. Pressing YES results in copying being carried out at the indicated position.

Tip
After positioning to the TO position via Locators/Markers, Project Top/End etc. (without the Dial) the copy is then no longer visible. By turning the Dial one grid step forwards and backwards, the copy appears exactly at the desired time position.

Application:
- Precise graphical positioning.
- Editing work without using the Mouse.

Word Clock
For Audio transfers between two (or more) digital devices, it is preferable to use one common System clock instead of the individual →Clocks for synchronizing the unit's →Sample Rates. This prevents digital clicks, glitches and Jitters. Generally a separate Word Clock Generator is used for this. Among the VS Models, only the VS-2480 (→Master Clock) provides a W. input in the form of the common BNC connection. To simply connect two digital devices, however, the self clocking operation mode that is contained within the data flow is sufficient (→Recording of digital Signals). In exceptional cases, numerous devices can be connected one after the other (Daisy Chain). The →ADA-7000 and →AE-7000 Interfaces can be simultaneously operated as Wordclock Master and Slave.

Word Length

Word Length
When digitalizing an Audio Signal for the Volume, the W. determines the →Resolution through the number of bits. In the VS, this is specified with the →Record Mode parameter when creating a new project →Project New.

White Noise
The entire statically distributed Sine Waves in all audible frequency ranges is referred to as Noise. In W., the spectral intensity in all frequency bands is constant and in comparison to →Pink Noise W. sounds therefore more accentuated in the higher frequencies. The →Generator/Oscillator of the VS-24xx is able to create white and pink Noise.

Write
Recording the Parameter Data in the →Automix and →Phrase Sequence is referred to as "Write". This is achieved independently of the Audio data already present as so called →Events in the Automix Tracks.

Writing Parameter
Within →Automix, W. specifies the mixer parameters, which are to be automated (→3. AUTOMIX SETUP Menu (p. 21)). "Level" and "Panorama" are default settings for normal use. Parameters which go beyond these firstly need to be activated: EQ; Mute; aux send-Level; →Insert settings and Surround. The desired section is then to be chosen. Generally this concerns the Parameters of the Track Mixer. For external effect devices or Sound modules synchronized via →MIDI Sequencer, the Automix parameters are to be set in the Input Mixer. In detail, the Parameters are:

Input- and Track Mixer:

- Volume (5 Parameters):
  Level; Pan; Offset Level/Pan; Mute
- Aux send (16 Parameters):
  Switch and Level
- Surround (4 Parameters):
  Pan; Depth; Center; LFE (Subw.Lv)
- EQ (12 Parameters):
  Switch;
  Shelving: (Gain, Freq);
  Peaking: (Gain, Freq, Q)
- Insert (16 Parameters in the VS-2400)
  pro Effect: Send Lvl; Return Lvl
- RSS (2 Parameters)
  Pan Mode (RSS in/out); RSS Pan

FX-Return:

- Volume (3 Parameters):
  Level; Balance; Mute
- Aux send (16 Parameters):
  Switch and Level
- Surround (4 Parameters):
  Pan; Depth; Center; LFE (Subw.Lv)

Effect

- Patch

Master

- Level (2 Parameters)
  Level; Balance

Aux Send and Direct Path

- (respectively 3 Parameters)
  Level; Balance; Position (Pre/Post)

V-Link
(7 Parameters)
Bank; Clip; Brightness; Color (Cb / Cr)

X

XLR
The →Balanced connection standard XLR is used for the connection of microphones, balanced →Line Level signals and for transferring digital Audio Data in the →AES/EBU format (→Cable).

XV-5080
The Sound/Sampling module "Roland XV-5080" provides 8 individual Outs for single outputting of sounds via →R-Bus.

Z

Zero
Using the →Transport Buttons ZERO or clicking on the [◄◄] in the VGA moves to the →Time Position 00:00:00:00:00.

Zip Drive
Although no longer supported, the →SCSI connector enabled a →System Update via the Z. in the VS-2480. Due to the limited memory capacity and the low data transfer, Audio recordings are only suggested in exceptional cases.

Zoom
Both the →Playlist and the →Waveform display can be increased or decreased in the VS. This is possible in both vertical (Tracks or Amplitude) and horizontal (Time) directions:

VGA:	Right click in the Wave Window →	ZOOM ↑↓ ZOOM ↔
VGA:	Click in "+"or "-" under the scroll bar	
VS:	SHIFT + Cursor ◄►, ▲▼	
Button:	SHIFT + ◄►, ▲▼ →Keyboard PS/2	

Appendix

Comparison Chart of Functions

Feature	VS-2480	VS-2400	VS-2000
19 " Rack mount	-	x	x
VGA, Mouse, Keyboard	x	x	optional
Recording Tracks simultaneously	16	16*	8
Channel Strip	x	-	x
Compressor/Expander simultaneously	x	-	-
Effect-Boards max.	4	2	3
Inputs with Phantom Power / without	8 / 8	8 / 0	8 / 0
Faders	17 x Motor	13 x Motor	17
Additional Filter (HP, LP, BP, Notch)	x	-	-
Generator/Oscillator and Analyzer	x	x	-
Harmony	-	-	x
Jog Wheel	x	-	-
JUMP Function	x	x	-
Channels Input Mixer	24	16	10
Headphone amps	2	1	1
LCD	medium	small	small
Locators	100	100	60
Mixer Channels in total	69	52	40
Monitor and Phones independently	x	-	x

Feature	VS-2480	VS-2400	VS-2000
Patchbay	x	x	-
R-Bus Patching In - Out	x	-	-
Record Modes 24/16/MTP/MT1/MT2	x	x	16/24
Rhythm Track (including SMF-Import)	-	-	x
S/PDIF Patching In - Out	x	-	-
Sample Rates 96/64/48/44,1/32	x	x	44,1
Sampler Function	x	-	-
SCSI	x	-	-
SMTP In	x	-	-
Surround	x	x	-
Track Direct Outs	x	x	-
Tracks	24	24	18
Tuner	-	-	x
USB	-	-	x
V-Link	-	x	x
Variable Effect Bus Assign	x	x	-
Vary Pitch	x	x	-
VGA Display "Multi Channel View"	x	x	-
Word Clock In	x	-	-

* 8 Recording Tracks on Board, 16 via. R-Bus

Important parameters and operations

Analog Inputs	15		Mixer	128
Arming Tracks	17		MONITOR OUT	132
AUX Send	27		Mouse	132
Backup	29		MT1, MT2	133
Balance	29		MTP	133
CD Backup	34		New Project	135
CD Burn	35		Operation Display	138
CD Marker	40		Outputs	139
CD Recover	40		PAGE	140
CD-R Write	41		Panorama	141
CH EDIT	42		Param1, 2 …	142
Clipping	45		Partition	143
Current Project	50		Peak Indicator	145
Current Time	50		Phantom Power	145
Cursor	50		Phantom Sw	145
Cursor Buttons	51		PHONES	146
Data Dial	51		Play	158
Defragmentation	52		Play Status	158
Drive	58		Playlist	158
Dual-Function Buttons	59		PLY	162
ENTER/YES	68		Position Line	162
EXIT/NO	71		Project	165
F Buttons	76		Project Backup	166
Fader	77		Project List	167
Fader Buttons	77		Project Load	167
Finalize	78		Project New	168
Function Buttons	81		Project Recover	169
Guitar Hi-Z	88		Project Store	169
Headphones	92		Record Mode	175
HOME	93		Record Standby	176
Initialize Mixer	96		Recording	176
INPUT	96		Safety in Operation	199
INPUT Level Meter	97		Sample Rate	199
Input Mixer	97		Save	200
Jogwheel	99		Send/Return Effect	204
LCD Display	102		Sensitivity Knob	205
Level Meter	103		Shift	206
Level Setting Recording	104		Store	211
List	105		Store Current?	212
Loop Effect	108		Switch off	215
M16	109		Timeline	225
M24	109		TR Mix	226
Mark	109		Track	226
Master Fader	112		Track Mixer	228
MASTER OUT	112		Track Status	229
Mastering Room	115		Transport Buttons	230
Master Tracks	118		Undo	231
Mix	125			

Bibliography

Roland: Owner's Manuals "VS-2000", "VS-2400" and "VS-2480"
Görne, Thomas: "Mikrofone in Theorie und Praxis", Elektor-Verlag
Krebs, Julius: "Das virtuelle Tonstudio, Roland VS-1680 und 1880", PlancTon Verlag
Dickreiter, Michael: "Handbuch der Tonstudiotechnik", KG Saur Verlag
Ackerstaff, Hans-Jürgen: "Mikrofone", Musik Produktiv
Mücher, Michael: "Fachwörterbuch der Fernsehstudio- und Videotechnik", BET Verlag
Pieper, Frank: "Das Effecte Praxisbuch", GC Carstensen Verlag
Bremm, Peter: "Das digitale Tonstudio", PPV Medien
Enders, Bernd: "Lexikon Musikelektronik", Goldmann/Schott Verlag
Katz, Bob: "Mastering Audio, the Art and the Science", Focal Press
"Production Partner" and "Keyboards", Music-Media Verlag
"studio magazin", Studio Presse Verlag
"Keys", PPV Verlag

Websites

Roland Deutschland:	www.rolandmusik.de
Roland USA:	www.rolandus.com
Roland Japan:	www.roland.co.jp
Edirol:	www.edirol.com
Antares:	www.antarestech.com
Universal Audio:	www.uaudio.com
IK Multimedia:	www.t-racks.com
Massenburg:	www.massenburg.com
McDSP:	www.mcdsp.com
T.C.:	www.tcelectronic.com
SoundToys:	www.soundtoys.com
PlancTon:	www.plancton.de
VS Planet:	www.vsplanet.com
VS-Forum Germany:	www.vs-forum.de

Das virtuelle Tonstudio
Roland VS-1680 und 1880

Practical Reference Book for Roland's VS-1680/1880/1824

A4 Format, Brochure, 260 pages
ISBN 3-00-005933-4   Price: 29,90 $

This reference book, available in German, describes necessary work processes for recording and editing music with the Roland VS-1680 to VS-1824 written in a style, which is easy to understand. Numerous tips for practical use regarding the topics of Recording, Editing, Synchronization, Mastering, Computer involvement etc. top this book off to make it the ideal companion for working with the VS.

The accompanying CD-Rom was recorded by the author using the VS-1680 and contains a Demo Song ("The Way to Start"), an interactive teaching song about automatic editing ("Fire in the G") and examples of applications for Microphone modelling and also Effects for the Monitor Mix.

VS-Tutorial
DVD

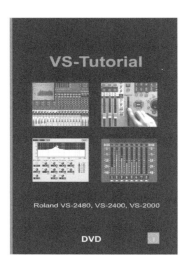

Tutorial DVD for Roland's VS-2480/2400/2000

More than 2 hours of explanations in English
ISBN 3-00-005933-4   Price: 29,90 $

This DVD is the ideal tutorial for successful production with the Roland VS 2xxx Recorders. The following work stages of audio production are shown in depth:

- Recording
- Editing
- Mixing
- Mastering
- CD-Burning

The display of the menus shown as they are in the VGA together with the camera position from the user's perspective and the additional animated graphics contribute to making this an incredibly practical learning tool. Thanks to detailed descriptions of the most important procedural steps and numerous practical tips, this DVD is suited to beginners as much as it is to professionals.

"The VS Encyclopedia" is also available in German as "Handbuch der VS-Recorder".
Plancton Verlag Berlin www.plancton.de

Memo

Memo

Memo